Primary Science

Teaching Theory and Practice

Fifth edition

Achieving QTS

Meeting the Professional Standards Framework

Primary Science

Teaching Theory and Practice

Fifth edition

John Sharp • Graham Peacock • Rob Johnsey
Shirley Simon • Robin Smith
Alan Cross • Diane Harris

LearningMatters

Acknowledgements
The authors wish to thank Jill Jesson for her work on updating Chapter 10 on Using ICT in science for the 4th edition.

Chapter 3: graphs reproduced by Joshua Harris.
Chapter 4: illustrations by Julie Bateman, aged 10.
Chapter 8: class organisation and interactive display diagrams by Joel Morris.
Chapter 9: explanatory pictures reproduced by Joshua, Laura, Jacob and Barnaby Harris. Concept cartoon reproduced from *Concept Cartoons in Science Education* (Naylor and Keogh, 2000) by kind permission of Millgate House Publishers.
Chapter 10: screen shot of circuit diagram from *Interfact: Electricity and Magnetism CD-ROM* (1997) by kind permission of Two-Can Publishing.
Chapter 11: health and safety diagrams by Joel Morris.

First published in 2000 by Learning Matters Ltd
Reprinted in 2001
Second edition published in 2002
Reprinted in 2002
Reprinted in 2003 (twice)
Reprinted in 2004
Reprinted in 2005 (twice)
Third edition published in 2007
Reprinted in 2007
Reprinted in 2008
Fourth edition published in 2009
Reprinted in 2009
Fifth edition published in 2011

British Library Cataloguing in Publication Data
A CIP record for this book is available from the British Library.

Printed book ISBN: 978 0 85725 086 5
Adobe ebook ISBN: ISBN: 978 0 85725 088 9
EPUB ebook: ISBN: 978 0 85725 087 2
Kindle ISBN: 978 0 85725 089 6

Cover design by Toucan
Text design by Bob Rowinski at Code 5 Design Associates Ltd
Project management by Deer Park Productions
Typeset by PDQ Typesetting Ltd, Newcastle-under-Lyme
Printed and bound in Great Britain by MPG Books Group, Bodmin, Cornwall

Learning Matters
20 Cathedral Yard
Exeter EX1 1HB
Tel: 01392 215560
info@learningmatters.co.uk
www.learningmatters.co.uk

Contents

1
Introduction

About this book

This book has been written to address the needs of all primary trainees on all courses of initial teacher training in England and other parts of the UK where a secure knowledge and understanding of how to teach science is required for the award of Qualified Teacher Status (QTS) or its equivalent. This book will also be found useful by Newly Qualified Teachers (NQTs), mentors, curriculum co-ordinators and other professionals working in education who have identified aspects of their science practice which require attention or who need a single resource to recommend to colleagues.

Features of the main chapters of this book include:

- clear links with the Professional Standards for Teachers (QTS level);
- references to required science knowledge and understanding;
- research summaries that give insights into how the theory of science teaching has developed, including seminal studies on children's ideas about science and how their understanding develops;
- examples of practice in the classroom to illustrate important points;
- suggestions for embedding ICT in your practice;
- reminders of how planning for teaching science fits in with the bigger picture, across the curriculum and with other aspects of school life;
- reflective tasks and practical activities for you to undertake;
- a summary of key learning points;
- references to key texts and suggestions for further reading.

For those undertaking credits for a Masters Degree, we have included suggestions for further work and extended study at the end of each chapter in a section called 'M-Level Extension'. The book also contains a glossary of terms.

What is primary science and why is it taught?

Primary science means different things to different people. Considerable and often heated debate in recent years has revolved around the portrayal of primary science as *product*, in which scientific knowledge is arrived at by objective methods capable of yielding accepted concepts, or *process*, in which scientific knowledge is arrived at by subjective acts of individual discovery driven by the development of scientific skills. In terms of the nature, teaching and learning of primary science, both have something to offer, and clearly an appropriate balance between the two is required. Primary science is perhaps best regarded, therefore, as an intellectual, practical, creative and social endeavour which seeks to help children to better understand and make sense of the world in which they live. Primary science should involve children in thinking and working in particular ways in the pursuit of reliable knowledge. While practical work undoubtedly contributes towards securing children's interest, curiosity and progress in science, children's scientific knowledge and understanding cannot always be developed through practical work alone. Just as the methods of science need to be taught explicitly, so too does the scientific knowledge and understanding implicit in scientific activities and their outcomes. As the National Curriculum Order for Science stated (DfEE/QCA, 1999):

> *Science stimulates and excites children's curiosity about phenomena and events in the world around them. It also satisfies this curiosity with knowledge. Because science links direct practical experience with ideas, it can engage learners at many levels. Scientific method is about developing and evaluating explanations through experimental evidence and modelling. This is a spur to critical and creative thought. Through science, children understand how major scientific ideas contribute to technological change – impacting on industry, business and medicine and improving quality of life. Children recognise the cultural significance of science and trace its worldwide development. They learn to question and discuss science based issues that may affect their own lives, the direction of society and the future of the world.*

Teachers (and trainees) are, of course, instrumental in developing children's scientific ideas and practical abilities and in fostering positive attitudes towards science. If you wish to find out more about primary science in general, you are directed towards the further reading and references sections included at the end of this introduction.

Professional Standards for QTS for Primary Science

Professional Standards for QTS (DfES/TDA, 2007) deals with the subject, pedagogical and professional knowledge and understanding required by trainees to secure children's progress in science. This book refers mostly to the pedagogical and professional requirements (see Peacock *et al.*, 2011 for subject knowledge and understanding). In summary, by the end of all courses of initial teacher training, all trainees are expected to know and understand:

- the reasons why it is important for all children to learn science and the value of engaging all children's interest in science;
- the nature of scientific understanding;
- key aspects of science underpinning children's progress in acquiring scientific knowledge, understanding and skills and how progress is recognised and encouraged;
- methods of developing children's scientific knowledge, understanding and skills;
- ways of organising and managing science in the classroom;
- assessing and evaluating science teaching and learning;
- the importance of health and safety;
- the benefits of using ICT in science.

Curriculum context

At the time of writing this edition, as at various other times in education, there are many possible changes to the curriculum and to the requirements for the subjects within it. The government that took office in May 2010 was planning a reform of

Early Years policy that will affect the Early Years Foundation Stage (EYFS). The Department for Education (DfE) was undertaking a review of the primary National Curriculum. Any changes to the National Curriculum or to the EYFS framework will be likely to have a knock-on effect on the assessment of children's attainment in the Early Years and at Key Stages 1 and 2, and there has already been much debate about this. This book includes some references to the statutory programmes of study for National Curriculum science, which maintained schools must follow until any new curriculum is in place, to the Early Learning Goals for children in the Early Years and to the systems of assessment that relate to these. In any transitional period, you will need to understand about what curriculum and assessment requirements were in place before the new arrangements and teachers you work with may retain in their practice elements of the earlier ways of working. You will certainly hear colleagues discussing the differences between curriculum initiatives and referring to former frameworks, for example the Primary National Strategy for Teaching Literacy and Mathematics, and to the QCA Scheme of Work for Science at Key Stages 1 and 2.

Because of these likely changes, we have focused in this book on giving you insights into the development of the theory behind the core areas of science teaching that you will need to inform your practice as you plan to promote the scientific understanding of the children you work with and to ensure that they develop effective investigative skills.

Early Years Foundation Stage

In September 2008, the Early Years Foundation Stage (DCSF, 2008) framework became statutory for all Early Years care and education providers who are registered with Ofsted. This framework applies to children from birth until they are five years old and therefore ensures that everyone involved in the care and education of the children is working towards common principles. Included in the experience it specifies are aspects of science.

Science in the National Curriculum

Science in the National Curriculum (DfEE/QCA, 1999) is organised on the basis of four key stages. Key Stage 1 for five- to seven-year-olds (Years 1 and 2) and Key Stage 2 for seven- to eleven-year-olds (Years 3 to 6) are for primary. The components of each key stage include programmes of study, which set out the science that children should be taught, attainment targets, which set out the science that children should know and be able to do, and level descriptions, which outline what children working at a particular level should be able to demonstrate. Science in the National Curriculum is a minimum statutory requirement. Since its introduction in 1989 it has been significantly revised three times. A brief summary of the programmes of study is presented as follows:

- Sc1: Scientific enquiry (Ideas and evidence in science; Investigative skills);
- Sc2: Life processes and living things (Life processes; Humans and other animals; Green plants; Variation and classification; Living things in their environment);
- Sc3: Materials and their properties (Grouping materials; Changing materials; Separating mixtures of materials – Key Stage 2 only);

- Sc4: Physical processes (Electricity; Forces and motion; Light and sound; The Earth and beyond – Key Stage 2 only).

Science in the National Curriculum also presents some information on the contexts in which primary science should be taught, links to other subjects, technological application, health and safety, and the use of ICT.

Use of an exemplar Scheme of Work for Science at Key Stages 1 and 2 (QCA/DfEE, 1998, with amendments 2000) was entirely optional. Designed to help implement Science in the National Curriculum, many schools adapted it for their own needs. The scheme was presented as a series of units which attempted to provide continuity and progression in primary science provision between Years 1 and 6. Guidance was offered on:

- the nature and place of each unit;
- how each unit builds on previous units;
- technical scientific vocabulary;
- resources;
- expectations;
- teaching activities;
- teaching outcomes;
- health and safety;
- the use of ICT.

You may find that individual teachers and schools that you work with are still using whole units or adapted parts of this scheme of work in their planning.

REFERENCES REFERENCES **REFERENCES** REFERENCES REFERENCES

DCSF (2008) *Statutory Framework for the Early Years Foundation Stage.* London: DCSF.

DfEE/QCA (1999) *Science: the National Curriculum for England*. London: HMSO.

DfES/TDA (2007*) Professional Standards for Qualified Teacher Status*. London: TDA.

Peacock, G., Sharp, J., Johnsey, R. and Wright, D. (2011) *Primary Science: Knowledge and Understanding*. Exeter: Learning Matters.

QCA/DfEE (1998, amended 2000) *Science: a Scheme of Work for Key Stages 1 and 2*. London: QCA.

FURTHER READING FURTHER READING **FURTHER READING**

Arthur, J., Grainger, T. and Wray, D. (2006) *Learning to Teach in the Primary School*. Oxford: Routledge.

Harlen, W. (2004) *The Teaching of Science in Primary Schools*. London: Fulton.

Hollins, M. and Whitby, V. (2001) *Progression in Primary Science: a Guide to the Nature and Practice of Science in Key Stages 1 and 2*. London: Fulton.

Roden, J. (2005) *Reflective Reader: Primary Science*. Exeter: Learning Matters.

2
The nature of scientific understanding

Introduction

This chapter discusses the very nature of scientific understanding and explores the implications for teaching science in schools. All of us behave in different ways as we learn more about the world around us. Often, however, how and what we learn leads to our own 'personal' understanding rather than that shared and accepted by the scientific community. While not everyone will become a professional scientist, those who use more scientific methods are more likely to have a more realistic understanding of how things are, unlike those who depend on hearsay or make inaccurate observations and poor interpretations.

The nature of scientific understanding

Science ... is the source of explanations about how and why things happen in the world around us ... [It should be seen] not as a set of facts to be learnt but as a series of explanations which the community of scientists currently considers to be best. (Watt, 1999)

It is easy to see why science has gained the reputation that it had in the past. In order to survive an often hostile world, it has been very important to establish how that world behaves and to predict what will happen next. The notion that science can provide watertight explanations and reasons is one we would like to believe in. Unfortunately, science cannot always provide clear-cut answers to everything, although many would like to believe so. In fact, the methods employed in science, and the body of knowledge which has been accumulated, provide only the best explanations we have so far, based on the evidence gathered and the interpretations put on that evidence.

It is true to say that there is some science that we are very sure about, largely because all the evidence collected over a long time points towards its validity. Thus, we are fairly sure that a force due to gravity will always pull an object which is close to the Earth downwards, and plants need a source of light in order to grow healthily. However, a scientist would say that if one day we found evidence that things do not always fall towards the Earth, we should then be prepared to change our views about how gravity works.

The ideas that are commonly accepted by the scientific community form the knowledge and understanding part of any educational curriculum and provide the ideas that scientists use to build new concepts and theories. While we should be ready to consider and reinterpret new evidence, we have to believe in some things or we may never leave our own homes for fear of floating off into outer space! At the same time, however, we must realise (and make others realise, too) that many ideas in science can never actually be proven but they can certainly be falsified.

IN THE CLASSROOM

Some teachers in a primary school noticed that often their children would be noisier and more agitated on windy days. The children came to school across a windy playground where the leaves themselves seemed to be swept up in frenzied excitement. Over a period of time, other teachers in the school made the same observation. Some teachers in the school claimed that their own classrooms were always calm despite the weather conditions outside. Observations at playtime on windy days, however, showed that the children from these classes were also particularly excited outside of their own classroom.

The group of teachers talked to colleagues from other schools and also read articles which supported their ideas about children's behaviour on windy days. The teachers formed a theory concerning the children's behaviour which was supported by keen observation and the collection of a range of evidence. The theory built up over a period of time and in some teachers' minds became fact.

There are a number of points about the nature of scientific understanding which this story can illustrate.

- The teachers had clarified their ideas about children's behaviour and windy days by making a general statement based on their initial experiences. *Scientific understanding is based on previously accumulated knowledge, which may be expressed in terms of generalisations.*
- Over a period of time, they checked their ideas against new evidence and found them to be consistent with this evidence. *The more evidence that supports an idea, the more we might accept it as valid.*
- However, even now, they cannot be sure that their ideas provide the best explanation because future observations may disprove them (in which case new, modified ideas may emerge). *Scientific ideas are often tentative.*
- As the teachers made more observations and developed ideas about why the children behaved

as they did, a theory emerged which could be tested. As long as the theory was supported by evidence, it could be usefully employed by some teachers to predict their children's behaviour and adapt the day's work to suit this. *A successful theory will enable successful predictions to be made.*

- An educational researcher might have been able to take a more scientific approach to testing this theory by making more reliable, consistent and repeated observations. Interpretation of this evidence might have produced a more sophisticated theory, which linked weather conditions to the general behaviour of the children, or disproved the theory altogether. *The quality of scientific knowledge and understanding is dependent on the quality of the scientific skills used to gather evidence and interpret it.*

The characteristics of scientific understanding

Harlen (2000, p. 17) describes four characteristics of a modern view of science.

- Science activity is about understanding.
- Science activity is a human endeavour.
- Science ideas are often tentative.
- Science ideas must always be evaluated against what happens in the real world.

Understanding in science involves providing explanations and searching for relationships between events, based on sound evidence. The evidence, however, is gathered and interpreted by human beings who, as we know, don't always get it right. Scientific ideas, then, are not a set of abstract rules set out in a textbook but rather a collection of (in some instances rather shaky) ideas set out by people who have interpreted what they have observed in their own personal way. One scientist may be mistaken or may have made an incorrect interpretation of the evidence. The views of a community of scientists who have critically checked each other's findings is much more reliable.

The tentative nature of all scientific ideas can be illustrated by putting ourselves in the place of those who believed the Earth was flat. This view of the world made a great deal of sense to most people who very rarely ventured far from home and could see with their own eyes an approximately flat landscape. This view of the world was acceptable and worked for those people on a day-to-day basis. Only when travellers and explorers confirmed that there was no 'edge' to the world and people began to notice that the masts of ships appeared first over the horizon was this view challenged. The evidence simply did not fit. Nowadays, we have even more evidence that the Earth is almost spherical in photographs from outer space. If, however, new photographs began to show that the Earth was doughnut shaped (unlikely of course), we would have to change our minds on the subject and develop new ideas!

Moral and ethical influences on scientific understanding

There are often more powerful influences on what we believe and perceive of the world besides the interpretation of scientific evidence. Galileo was persecuted by the Church for suggesting a theory which put the Sun at the centre of our Solar System with our rather insignificant Earth orbiting it. The Church had taught an Earth-centred view of the Universe and felt compelled to suppress the scientific evidence that Galileo produced. Modern debates over scientific developments can be influenced as much by politics and prejudice as by hard scientific evidence. Interpretations of evidence can be purposely or inadvertently misleading as a result of the interpreter's moral or ethical stance. Commercial pressures can encourage unscrupulous groups to misconstrue evidence or even falsify it. In short, scientific progress can be hindered and distorted in a variety of ways.

THE BIGGER PICTURE   THE BIGGER PICTURE   THE BIGGER PICTURE

When you are planning for a unit on The Earth and beyond with older primary children, consider including links to the history of when people believed that the Earth was flat, and setting a research task on the discovery and reclassification of Pluto. These are practical examples of the changing nature of scientific understanding and may help children to assimilate this important point about science as a subject.

If children are to cope in a modern technological society, they must come to appreciate the ways in which scientific knowledge is generated and the ways in which scientific ideas have been developed as a result of human endeavour. While a science curriculum will define a body of knowledge that is apparently fixed and irrefutable, teachers must issue a warning that this is not necessarily so. This can often be achieved through a discussion of the tentative nature of the children's findings in their own investigations. The children themselves know just how shaky some of their experimental results are and should understand that all scientists will have suffered the same kind of problems.

Science and design technology

Sometimes it is important to know what science is not, in order to understand what it is. To many people, science and technology are spoken of in the same breath as though they were one thing, but this is to misunderstand an important fact. Science is to do with finding out about how the world around us works and how it behaves. This is achieved through formulating and testing theories and models by collecting and interpreting evidence. The purpose is to generate reliable knowledge. Design technology, on the other hand, is primarily concerned with solving problems and satisfying human needs. It draws heavily on the knowledge gained in science and other disciplines, but it uses this knowledge to achieve a different end.

A science curriculum needs examples of technology to help children further understand the science. Children need to understand the implications of using science ideas. When debates rage over such things as genetically modified crops or the use of nuclear power stations, the argument is with the technologists – those who are employing the knowledge gained through scientific pursuits. The pure scientists only supply the knowledge they have gained.

RESEARCH SUMMARY RESEARCH SUMMARY RESEARCH SUMMARY **RESEARCH SUMMARY**

Research by McCormick (1999) into the way children carry out design technology tasks focused on the way in which the children applied their knowledge in a practical problem-solving situation. He found that while children might be working on a product with a number of scientific features, their discussions about it were not made in an abstract scientific language. He describes how two children were designing and making a collecting box in which a coin made a model bird move. Their discussions involved a technological language which was closely related to the actual model rather than the scientific or mathematical language they might have employed in a formal science lesson.

McCormick concluded that the information needed in the practical context was subtly different to that gained in the more theoretical science lesson. He called this knowledge 'device knowledge'. This suggests something about the nature of science knowledge as taught in schools and points towards a need for greater links to be made between this type of knowledge and the device knowledge often used in the practical world.

Implications for science teaching

The science curriculum should provide young people with an understanding of some key ideas-about-science, that is, ideas about the ways in which reliable knowledge of the natural world has been and is being obtained. (Millar and Osborne, 1998, Recommendation 6)

The National Curriculum for science in primary schools appears to be set out largely in terms of science facts and concepts. These are the 'big' science ideas about which we are fairly certain and which are deemed appropriate for primary-age children to understand. It is right that teachers teach these ideas as firmly held concepts but this should be done alongside stories about the sometimes tentative nature of modern science and about how some old ideas have been modified or replaced throughout history. At the same time, teachers will want to teach the skills of science enquiry to illustrate the way in which scientific ideas have been arrived at in the past and are arrived at today.

A SUMMARY OF **KEY POINTS**

> **Science is about understanding the world around us – technology is about using this knowledge to solve problems.**
> **Scientific understanding is based on accumulating and interpreting evidence for proposed theories or models.**
> **Science ideas are often tentative but the 'big ideas' are well established.**
> **Science is a result of human endeavour and thus scientific interpretations may be influenced by moral, ethical, political and commercial factors.**

M-LEVEL EXTENSION > > > > M-LEVEL EXTENSION > > > >

Look again at Harlen's (2000) four characteristics of a modern view of science that were discussed earlier in this chapter. Drawing on your study of reference texts and the suggestions for further reading, how would these four characteristics affect your planning for the children you are currently working with? What elements will you include so that primary children can develop a growing understanding of the nature of science?

REFERENCES REFERENCES **REFERENCES** REFERENCES REFERENCES

Harlen, W. (2000) *Teaching, Learning and Assessing Science 5–12*, 3rd edn. London: Paul Chapman.

McCormick, R. (1999) Capability lost and found. *Journal of Design and Technology Education*, 4(1), 5–14.

Millar, R. and Osborne, J. (1998) *Beyond 2000 – Science Education for the Future*. London: King's College.

Watt, D. (1999) Science: learning to explain how the world works, in Riley, J. and Prentice, R. (eds), *The Curriculum for 7–11 year olds*. London: Paul Chapman.

FURTHER READING FURTHER READING **FURTHER READING** FURTHER READING

Driver, R., Leach, J., Millar, R. and Scott, P. (1996) *Young People's Images of Science*. Buckingham: Open University Press. This book provides an insight into the nature of scientific understanding as a whole as well as how children's ideas about the nature of scientific understanding develop.

Ollerenshaw, C. and Ritchie, R. (1997) *Primary Science – Making it Work*. London: Fulton. This book provides a general introduction to teaching primary science, drawing heavily on the constructivist approach – one which has been adopted particularly by the science education community.

Peacock, G. A. (2002) *Teaching Science in Primary Schools*. London: Letts. A clear but brief introduction to the intricacies of teaching science in the primary school.

3
Processes and methods of scientific enquiry

Professional Standards for the award of QTS

Those awarded QTS must have a secure knowledge and understanding of science that enables them to teach effectively across the age and ability range for which they are trained. To be able to do this in the context of understanding the processes and methods of scientific enquiry, trainees should:

Q4 Communicate effectively with children, young people, colleagues, parents and carers.

Q7(a) Reflect on and improve their practice, and take responsibility for identifying and meeting their developing professional needs.

Q10 Have a knowledge and understanding of a range of teaching, learning and behaviour management strategies and know how to use and adapt them, including how to personalise learning and provide opportunities for all learners to achieve their potential.

Q12 Know a range of approaches to assessment, including the importance of formative assessment.

Q14 Have a secure knowledge and understanding of their subjects/curriculum areas and related pedagogy to enable them to teach effectively across the age and ability range for which they are trained.

Q22 Plan for progression across the age and ability range for which they are trained, designing effective learning sequences within lessons and across series of lessons and demonstrating secure subject/curriculum knowledge.

Q25 Teach lessons and sequences of lessons across the age and ability range for which they are trained in which they:

(a) use a range of teaching strategies and resources, including e-learning, taking practical account of diversity and promoting equality and inclusion;

(b) build on prior knowledge, develop concepts and processes, enable learners to apply new knowledge, understanding and skills, and meet learning objectives.

Q28 Support and guide learners to reflect on their learning, identify the progress they have made and identify their emerging learning needs.

The guidance accompanying these standards clarifies these requirements and you will find it helpful to read through the appropriate section of this guidance for further support.

Introduction

IN THE CLASSROOM

A class of six- and seven-year-olds were working in the small garden the school had created. The tasks for the day were planting and weeding but several children had become fascinated by the small creatures they found as they worked. One group looked under stones and discovered more. Some children had names for creatures they saw. Others pointed out how they moved. Their teacher had not planned for this to be the focus of the science but it fitted the school scheme of work well, since they were studying the variety of life and conditions for living things. She recognised its potential to develop scientific enquiry from the children's curiosity and interest. Before they went any further, she talked with them about being careful not to harm any creatures. Some children wanted to make homes for the ones they had found. The teacher asked what they thought these would have to be like in order to start them thinking about the types of environment which different animals needed. She anticipated helping them investigate the conditions which the small animals preferred, working with them to devise simple tests of light/dark and damp/dry.

The class had already been shown how to use a magnifying glass and viewers, so they would be able to make closer observations to check the claims they were making about how many legs each creature had. The teacher thought that drawing and maybe modelling might improve the observations. That would also help them compare different animals with pictures in books and she could introduce simple keys and some software the school had. Grouping pictures, sorting activities and games might help children get an idea of a simple classification. At present, some of them were using terms like worm and insect but there were lots of disagreements over what to call each new find. The teacher could see opportunities for language work in these debates, and there would be maths to apply in drawing charts or graphs if they found lots of things. She could also see some challenges in organising all this – not least with the two children who didn't want to have anything to do with 'creepy-crawlies'.

This scenario illustrates some opportunities to develop processes and methods used in scientific enquiry, not all of which are exclusive to science. Compare it with the way that the National Curriculum for Science (DfEE/QCA, 1999) describes what children should do at this stage.

During Key Stage 1 pupils observe, explore and ask questions about living things, materials and phenomena. They begin to work together to collect evidence to help them answer questions and to link this to simple scientific ideas. They evaluate evidence and consider whether tests or comparisons are fair. They use reference materials to find out more about scientific ideas. They share their ideas and communicate them in language, drawing, charts and tables. (p. 78)

When they build on this during Key Stage 2, one sign of progress is that children carry out more scientific investigations.

Investigations and other practical work

Investigations are an important part of science in primary schools. They develop children's understanding of scientific procedures and concepts, their knowledge and their skills.

Investigations are not simply practical work. Practical activities may be used for various reasons, for instance to illustrate a scientific concept, to teach a specific skill or to foster observation. These are all worthwhile. However, it is important to be clear about the purpose of any activity – for example, giving children a detailed set of instructions to follow so they reach a predetermined result may be appropriate if we are aiming to teach skills or to reinforce ideas. However, this is not the same as children carrying out investigations where the aim is to allow them to use and to develop concepts, skills and procedural understanding. When they are investigating, children should have some responsibility for deciding what to do and how to do it. They should be encouraged to think for themselves, to use their knowledge and skills, and to extend their understanding.

Characteristic features of investigations

Children may have to work out exactly what question they are trying to answer and turn it into a form that can be investigated.

They should be able to use investigative procedures, such as:

- raising questions;
- identifying and controlling variables;
- planning;
- observing/measuring;
- analysing data;
- interpreting results;
- deciding how far their findings answer their original question;
- evaluating their work.

A whole investigation may include all these features. However, we often support children by organising investigations so that they can concentrate on a few aspects. We also have to teach them the skills and knowledge that they need in order to carry out investigations. Although children are taking responsibility for their investigation, teachers have an active part to play – they are not simply providing hands-on experiences nor adopting a discovery approach to science learning.

PRACTICAL TASK PRACTICAL TASK **PRACTICAL TASK** PRACTICAL TASK

Consider a common practical activity in primary science – bouncing balls.

1. How might you organise the activity if you want to teach a skill – e.g. careful observation by young children or accurate measurement by older children?

---

2. How might you organise the activity if you want to ensure all the children get a good set of results that can be used to see patterns in results and a relationship between the height of drop and height bounced?

3. How might you organise the activity if you want to extend children's skills at planning a fair test where the key variables are tested and others are controlled?

4. How might you organise the activity if you want children to carry out an investigation of their own?

Make notes or draft a lesson plan for each of these different objectives.

Now that we have an overall picture of investigations, we will look at some of the features in more detail.

Raising and framing questions

Children ask lots of questions although, sadly, many are discouraged even before they arrive at school. So, teachers may have to encourage them to raise questions through devices such as unusual displays, boxes or boards where children can post questions. When starting new topics, lists of potential questions can be discussed and used in planning the work with the children. There should be a school climate where questions are treated seriously and teachers should be role models. They will certainly need to help children pose questions that are framed so that they can investigate them and find something out. That is a crucial step in any research. The skill can be scaffolded with class discussion to refine questions posed by the children or teacher, with support for small groups working on their own ideas, through direct suggestions and through active redrafting of written questions. As they gain experience and confidence children can be challenged more:

- Are they making use of what they already know to pose their question and predict what might happen?
- Have they thought about different ways their question could be investigated?
- What skills have they learnt already that would help them with the investigation?

Fair testing and other investigations

Primary teachers have been very active in teaching children about the need for fair tests. These are investigations in which the key independent and dependent variables are systematically changed and measured while the others are controlled so they do not change. The independent variable is the one that will be changed intentionally to see what effect it has on the dependent variable which will be measured. Variables are those things which change or vary in an investigation. How do children and teachers know what they are likely to be in any specific case, for example when testing materials to see which is most effective for keeping someone warm? Well, their previous experiences of clothing and the scientific knowledge they have about materials will suggest some factors. The level of their understanding about how heat travels can make a difference to what factors they see as relevant. Children may need lots of help to work out which variable they are going to control, i.e. keep the same, and which they will change. By trying out their ideas practically they may find other things which affect the results. Teachers need a good understanding themselves of the skills and ideas that the tasks may demand – for instance, being able to use a thermometer and knowing the difference between heat and temperature. They have to be aware of practical difficulties and safety issues – can the water be hot enough for the children to get good results? And they also need to be alert to what the children are thinking – for example, do they expect water to go on cooling to zero rather than room temperature? Do they interpret 'fair' as meaning everybody has a go?

RESEARCH SUMMARY RESEARCH SUMMARY RESEARCH SUMMARY **RESEARCH SUMMARY**

Joyce *et al*. (2005) describe a study with five-to-seven-year-olds to help them develop their ability to control variables and keep track of them when conducting a fair test, based on running small cars down ramps. They used strategies such as setting up two sets of equipment side by side, providing pairs of cars that could be selected to create a fair test and introducing coloured dots for recording the distance travelled. These reduced the demand on the children's memories by displaying more of the overall investigation simultaneously, making the effects of 'unfair' testing more obvious, easing data recording so repeat tests were more manageable and making patterns of results obvious so children could compare and reason about them more easily. They found that with these strategies even young children were able to identify a fair test and replicate the set-up of equipment for their investigations. The children showed their reasoning as they talked about what might make a fair or unfair test.

In many investigations, variables can be difficult to control or they may interact. Some people also argue that fair tests have been emphasised too much to the detriment of other valuable work. This is not an argument against teaching children how to carry out controlled tests, but rather one for giving them experience of other kinds of investigation and other contexts. This range of work is not easy to fit in when teachers feel that the time available for primary children to do extended science investigations has been squeezed by the National Curriculum and Primary National Strategies, so it is crucial to avoid repetition. This is true both within and across key stages.

Planning investigations

In order to plan their own investigations, children need to be able to think ahead, anticipating what they will do and what equipment they will need, how they will record their findings and what they will do with the results. This is demanding and they will need a lot of help. In the early stages, teachers will do most of the planning for the children but should aim to involve them and make them aware of the process. Even young children can be given responsibility for planning part of an investigation. In their play they imagine and organise their activities. As they collaborate in groups and extend their communication skills children become able to take on more of the planning themselves. For teachers in primary schools nowadays, there is a tension between fostering this independence and following the prescribed curriculum. Rachel Sparks Linfield (2007) draws on her experience across the age range in describing how the National Curriculum, and particularly end-of-key-stage assessments, have restricted opportunities for children to pursue extended investigations. She makes the argument for regaining time to enable them to investigate their questions.

When they have had some experience of science activities, children can do more systematic planning of their investigations. Many teachers have found planning sheets helpful to structure this and to support children. These can be provided through a computer, on paper or on a larger-scale board. A few key headings or questions direct children's attention to things they should think about before they start, for instance for a 'fair test' investigation.

- What are we trying to find out?
- What are we changing?
- What do we want to keep the same?
- What do we think will happen?
- What are we going to measure?
- How will we record our measurements?
- What equipment do we need to collect before we start?

THE BIGGER PICTURE   THE BIGGER PICTURE   THE BIGGER PICTURE

When you are planning investigations, consider using guided writing techniques from your literacy work to support the children's learning in science. For example, you may want to introduce them to an effective way of planning or recording an experiment through the use of writing frames, recording grids and other scaffolding methods. Using guided writing techniques in other parts of the curriculum like this reinforces the expectation that literacy skills are not just for use in English sessions and can support the less able children or those with additional or special needs to be more fully engaged with their work in science. For more on writing frames and other methods of scaffolding, see the companion book to this one *Primary English: Teaching Theory and Practice* (Learning Matters, 2011).

The headings should prompt children to think through the stages of an investigation one at a time. They can help them focus their attention, be organised and think ahead to consider alternatives or problems. They might be referred to as a reminder during their investigation. The structure can also reinforce children's understanding of the procedures involved in an investigation. However, we should take care that such aids do not become straitjackets. Planning sheets usually emphasise one sort of investigation, the fair test. They may also become a routine or even a ritual, taking time and attention away from the investigation itself.

So, we need to be clear about our purpose for requiring the children to plan – and to share the purpose with them.

- Is it so they are more likely to do an appropriate investigation for the question they posed?
- Is it because they might miss important observations?
- Is it so that they think about the variables that need to be controlled?
- Is it because they have to plan what measurements to take and get the equipment?
- Is our aim to increase their understanding of the scientific procedures in a whole investigation?

REFLECTIVE TASK
REFLECTIVE TASK

Science investigations can also serve a broader purpose across the curriculum: fostering children's creativity and their thinking skills.

Think about investigations you have done or might plan to do with children.

1. What can you do in order to encourage children to generate lots of possible ways of investigating a question?
2. How might children be helped to discuss one another's ideas and decide on the most appropriate ways forward?
3. How confident would you be at letting children follow up their own ideas, even if they looked like being unsuccessful? How do you decide whether to intervene?

For ideas, look at articles on creativity in issue 81 of *Primary Science Review*, and articles on questions in issues 83, 90 and 105.

Observation and measurement

Careful observation is an important skill to develop in science. Teachers can use devices such as viewing frames or magnifying glasses to help children look more closely. They can ask children to report on changes over time, for example when food colouring is dropped into water. Audiotapes and videotapes can be replayed so they can check their observations. Teachers' questions can be used to focus or extend children's observations as appropriate: sometimes it is best to draw their attention to relevant features and at other times to ask them what else they notice.

The questions and problems that lead to investigations are often the result of observations, especially of noticing surprising features or unusual events. Observing is not simply looking or listening but is a deliberate act influenced by our expectations and experience. As children gain more scientific knowledge they can be asked to use that to inform their observations. In investigations they should be drawing on their scientific ideas from the start. Observation features at later stages as well. Some kinds of investigation depend on lots of carefully recorded observations to see if there are any patterns, for example if they want to investigate food preferences of birds visiting a school bird table. Children have to be prepared to make these – they may have to get equipment, practise using it, and decide when to make and how to record their observations. Some investigations combine qualitative records and measurement, for instance if they trace the germination and growth of plants under various conditions or keep a record of the weather.

Measuring is a key feature of science, although not all scientific work by children will involve measurement. Children need to see the purpose for measuring in their investigations. They should be helped to think about what they are going to measure and how they can do so. If they are going to need to use measuring instruments do they already know how to? Have they been taught how to use them appropriately? Are they seeking to be accurate and are they aware of errors? Consider, for instance, children carrying out a very common activity, rolling vehicles they have made down slopes to see which runs fastest. Children often opt to time these using a digital watch or stopclock and report their results. However, the time delays in switching the clock on and off can be considerable, and they need to realise this and to consider what else they might measure.

In tests like this children are often expected to make repeated measurements. This is good practice but once again we have to be careful that they learn to do it with understanding.

Some of the equipment children will use during investigations may be met first in maths when they measure time, length, weight and volume. As they progress through simple comparison and use of non-standard measurements to standard measures, they acquire skills that they can use in science. They are also learning about accuracy and choice of the appropriate scale. Other equipment such as spring balances to measure forces or thermometers to measure temperatures may be introduced in science lessons. They will need to be taught how to use these with skill and understanding, for example beginning with learning to read a simple scale on sturdy thermometers, progressing to recognising the range they can measure using more sensitive instruments, handling them carefully, and reducing errors by waiting for the instruments to reach a steady reading. At National Curriculum level 3, they are expected to be able to use a range of simple equipment to measure quantities like mass and length; by level 5 they should be selecting apparatus, planning how to use it, making a series of measurements with appropriate precision, and beginning to repeat measurements.

EMBEDDING ICT EMBEDDING ICT **EMBEDDING ICT** EMBEDDING ICT

ICT has brought new aids such as dataloggers with sensors that children can use to make and record measurements. These devices and software can help them handle the data they produce, see patterns and relationships and communicate their results. The links between maths, science and ICT are most obvious in the area of data handling. See Chapter 10 for more information on using dataloggers.

Analysing data and interpreting results

Young children need lots of help to organise their data collection and recording. The simplest form of recording is to present results directly, for instance by making a display that shows plants grown in different conditions or by placing materials in order from strongest to weakest as they are tested. Children at all stages can use tape recorders, digital cameras or video recorders to capture observations; this could be especially useful to enable children with communication difficulties to focus on the science and achieve their potential.

Progression in this aspect of their work can be illustrated by reference to the National Curriculum levels. At level 1, children will be using pictures and simple charts. If drawn large, they can be displayed and discussed as the first step in interpreting data. Talk and writing are important for communicating findings at any stage and mathematics contributes other ways of presenting results that children need to apply. In the early stages, teachers may provide a simple tally chart or table but as they progress children should begin to plan their own. At level 2, they are expected to use simple tables where appropriate. If they are to progress, teachers need not only to introduce them to other ways of recording but also to help them see patterns in data that have been recorded. At level 3, they are expected to record their own observations in a variety of ways and to provide explanations for simple patterns in recorded measurements. To build on that, they need to learn about using different sorts of tables, begin to plot points in simple graphs and use these graphs to interpret patterns in their data. At level 5, children should be able to make decisions about the best way to record and present particular sorts of data.

EMBEDDING ICT EMBEDDING ICT **EMBEDDING ICT** EMBEDDING ICT

Software such as spreadsheets and databases may assist children to process lots of data and to try different ways of displaying it, including as tables, bar graphs, pictograms, line graphs and scattergrams, but you should ensure they are taught to use them with understanding. For example, they should realise that when there is a relationship between variables, this makes it appropriate to draw a line graph, but also when that is not appropriate.

Children can work at different levels on related investigations in one class, as in the following examples of bouncing balls. Those who drew the first graph (see Figure 3.1) were working at a lower level in science and in maths. The teacher had suggested they compare how high different balls bounced. They chose the balls and did the tests without a lot of help but they only made one measurement for each, using a metre rule. This sort of graph was familiar to them. The second graph (see Figure 3.2) was drawn by the most able children in the class, who practised first in order to measure the height of bounce as accurately as possible. They then fixed a scale on the wall, dropped the ball at 10cm intervals, made three readings for each and took the median value. They realised that a line graph could be drawn to show the relationship between the two heights. In a previous investigation of cooling, they had plotted points for temperatures and had learnt to draw a line of best fit. When they drew this one, they decided after discussion that they could take the line back to zero, even though when they had dropped the ball from less than 40cm they had not got reliable measurements.

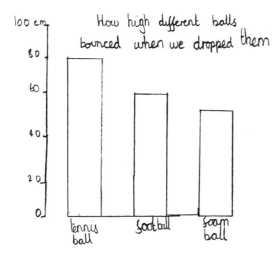

Figure 3.1 Bar graph

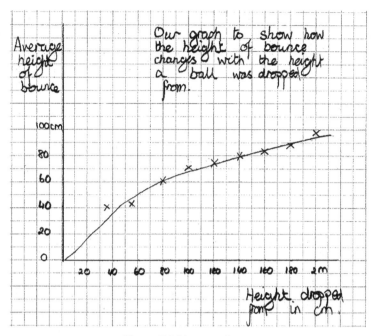

Figure 3.2 Line graph

As well as acquiring these skills and the procedural understanding to apply them, children at level 5 and above should be using their data to draw conclusions which are based on evidence and which relate to scientific ideas. This is the purpose of doing whole investigations, but the skills to enable this sometimes have to be taught separately and then put together. In some ways interpreting results is the hardest of these. It requires the children to scrutinise data for patterns, to relate the findings to their predictions and yet be open to unexpected results, and to draw on scientific ideas while their understanding is limited. There is plenty that teachers can do to help children with this as children progress through the primary school.

- Talk about patterns and regularities in everyday events, for example daily cycles or shadow lengths.
- Draw attention to phenomena which indicate relationships, for example how seedlings grow towards a window or how many snails are about on wet days.
- Display children's work so that patterns are apparent and discuss them, for example the shadows cast by a puppet figure at different distances from a lamp or the order in which parachutes of different sizes fell. Include questions in displays.
- Ask them to refer back to their predictions when they discuss the results of a test they did.
- Organise activities where measurement reveals relationships, for example between pulse rates and exercise.
- Provide tables or graphs with results for children to analyse and discuss, for example elastic bands stretched by loading masses. Introduce some with less tidy sets of results and ask for their explanations, for example the distance travelled by cotton reels powered by turns of an elastic band or the readings on a spring balance pulling increasing loads.
- When they report their findings, ask children to provide explanations and suggest reasons if any of their results do not fit the pattern.
- Encourage them to give reasons based on their science studies when they predict and offer explanations.

● Establish a climate where children can challenge one another's explanations, offer alternative interpretations and listen to one another.

Evaluating

Evaluating is valuable for improving investigations and developing procedural understanding. However, it should not become a ritual or a chore, nor do we want children to feel defensive about their work. Specific, focused questions will be more helpful at first rather than general enquiries about how they could make it better. For example, if children were recording something changing over time, ask how they decided when to take measurements, whether any measurements needed to be checked, what was difficult and what worked well. As children become more capable at planning investigations, they can look back at their plan to see how closely they followed it, what adjustments they had to make as they went along and how they would advise another group to tackle the same question. If several groups investigate similar questions, they can compare their findings and discuss why they may not have the same results. The time available for repeating and refining investigations is limited so we need to make the best use of any opportunities. Much of the evaluation can occur as children talk about the work they are doing. We also have to bear in mind that children often find science stimulating because they are actively engaged in practical work and trying out their own ideas. Teachers need to listen and watch them at work in order to judge whether a question will extend the children's understanding or be seen as a distraction. Here are some ways in which you can successfully develop children's thinking as they investigate.

RESEARCH SUMMARY RESEARCH SUMMARY RESEARCH SUMMARY **RESEARCH SUMMARY**

Keogh and Naylor (2003) identified some skills and capacities of Early Years teachers which foster scientific learning, such as:

● helping children observe what is important;

● being curious and willing to enquire themselves;

● having the capacity for surprise.

Moving up the age range, Stephen Lunn (2006) reported how a teacher with four- to six-year-olds and later with seven- to nine-year-olds built on their ideas and fostered imagination by using charts of 'what do we know' and 'what do we wonder'; these stimulated questioning, developed the children's confidence to offer ideas and modelled scientific thinking.

At the top of Key Stage 2, Richard Watkins (2005) researched the scientific reasoning of the children in his Year 6 class. He gave them statements reflecting either inductive or deductive reasoning to discuss in small groups. Most children could identify limitations of inductive reasoning as a basis for making predictions about objects not yet examined on the basis of information from those already examined.

Watkins teaches investigative work across the curriculum and also dedicates lessons specifically to teaching investigative skills, including questions of reliability and accuracy. Children are encouraged to think aloud and explain their reasoning. His account illustrates both the creation of a climate that fosters children's own ideas and the rigorous teaching of skills, enabling them to achieve progression in their scientific thinking.

A SUMMARY OF **KEY POINTS**

> Practical activities may be used for various reasons: to investigate a question or problem; to illustrate a scientific concept; to teach a specific skill; to foster observation.

> Investigations are an important part of science in primary schools. They develop children's understanding of scientific procedures and concepts, their knowledge and their skills.

> When they are investigating, children should be encouraged to think for themselves, to use their knowledge and skills, and to extend their understanding.

> Children should be taught to use and to understand investigative procedures such as framing a question in a form that can be investigated, identifying and controlling variables, planning, observing, measuring, analysing data, interpreting results, deciding how far their findings answer their original question and evaluating their work.

> Science investigations can be stimulating, enjoyable and challenging for children – and for their teachers!

M-LEVEL EXTENSION > > > > M-LEVEL EXTENSION > > > >

Think again about the practical activity discussed earlier in this chapter about bouncing balls. Reflect on the suggestions that you made for changing the activity to meet a range of different learning objectives and look again at the graphs produced by different groups of children undertaking similar investigations. How would you differentiate this activity for children of differing levels of ability? What about those children with specific learning difficulties or particular needs, for example children with English as an additional language? Would they require any particular support or additional resources? Remember that you may also have in the same group or class children who are able, gifted and talented in science. When you have made your decisions, try the same exercise based on the impromptu minibeast investigation described at the start of the chapter.

REFERENCES REFERENCES **REFERENCES** REFERENCES REFERENCES

DfEE/QCA (1999) *Science: the National Curriculum for England*. London: HMSO.

Joyce, C., Hipkins, R. and MacLeod, C. (2005) We need to get some more dots! *Primary Science Review*, 89, 11–14.

Keogh, B. and Naylor, S. (2003) Being a role model in the early years. *Primary Science Review*, 78, 7–9.

Linfield, R. S. (2007) Bringing investigation back to science. *Primary Science Review*, 100, 26–27.

Lunn, S. (2006) Working like real scientists. *Primary Science Review*, 94, 4–7.

Watkins, R. (2005) The pterodactyl and the crow. *Primary Science Review*, 89, 4–7.

FURTHER READING FURTHER READING **FURTHER READING** FURTHER READING

Goldsworthy, A. (2004) Acquiring scientific skills, in Sharp, J. (ed.) *Developing Primary Science*. Exeter: Learning Matters, 33–49. Goldsworthy illustrates many of the ideas discussed in this chapter. She refers to earlier projects and publications from the Association for Science Education (ASE), which remain valuable practical resources for teaching investigations, for example the AKSIS project and the book by Goldsworthy and Feasey, *Making Sense of Primary Investigations*, which is still available from the ASE.

The primary journal of the ASE, *Primary Science Review* (PSR), which since issue 101 is now called simply *Primary Science*, regularly contains articles illustrating children's investigations and ways that teachers foster their pupils' skills and reasoning. Some articles and special issues deal explicitly with these matters, for example:

PSR Issue 81 on Creativity includes an article that describes creative science in a nursery school by McFall and Macro on pages 7–10, and one by Ovens on pages 17–20 in which he illustrates ways to encourage creativity with Year 1 and Year 3 classes.

PSR Issue 90 on Forensic Science describes several investigations set in a 'crime-solving' context which developed children's enquiry skills, such as that by Ian Richardson 'The baker did it' on pages 4–6.

PSR Issue 100 on Children's Science and getting children to think scientifically.

Primary Science Issue 105 on Questioning.

4
Children's ideas

Introduction

To understand how children learn science, and the importance of children's ideas in the learning process, teachers need to have some knowledge of the nature of children's learning.

Piaget's theories of development and of how children construct their understanding as they interact with the world were an early influence in primary and nursery education. The importance of language, social interaction and instruction were subsequently acknowledged, with the work of Vygotsky being particularly influential. More recently, research into children's learning has sought a broader perspective that takes account of interactions between their developing conceptions, specific knowledge they have gained, their beliefs and ways of thinking and feeling.

Theories about children's learning

Many developments in science education in the past 30 years have been influenced by constructivist and social constructivist theories of learning. In essence, a constructivist view of learning suggests that learning involves an active process in which each learner is engaged in constructing meanings, whether from physical experiences, dialogue or texts. A constructivist perspective has underpinned many studies into children's ideas undertaken in different parts of the world. Highly influential projects included the New Zealand Learning in Science Project (LISP), the Children's Learning in Science Project (CLISP) and the primary Science Process and Concepts Exploration (SPACE) project. The following Research Summary provides an overview of some of this work.

RESEARCH SUMMARY RESEARCH SUMMARY RESEARCH SUMMARY **RESEARCH SUMMARY**

Following the seminal studies of LISP and CLISP in the 1980s, a wide range of studies across the world with children of all ages have shown that they develop meanings for many words used in science and views of the world that relate to ideas taught in science. The findings also show that children's ideas may be strongly held and often differ significantly from those held by scientists, that from the children's points of view their ideas are sensible and coherent, and that they often remain uninfluenced by science teaching or are influenced in unexpected ways.

The SPACE project began just before the National Curriculum. It carried out research, working with primary pupils over a range of scientific topics in classrooms, and built the findings into classroom guides for teachers. A concise account of its findings and legacy is given by Harlen (2007). She illustrates some of the alternative ideas held by children which were revealed in the study and may be recognised by many teachers. Although their ideas may be based on adequate evidence, it is often clear that children have thought about what is happening and that there is some reasoning based on observations.

Studying the characteristics and limitations of children's ideas gives us clues as to how to help children change their ideas to come closer to the scientific ones. (Harlen, 2007, p. 14)

Some more recent research has examined how children's ideas may change over several years, questioning the stability of individuals' ideas and the assumption that there is an orderly, sequential growth of concepts (Tytler and Peterson, 2005). In one study the researchers worked with and interviewed 12 primary school pupils each year from the age of five. They studied children's understanding of a number of topics, e.g. changes to materials and evaporation. They found that:

- children's explanations were very context dependent, and there was no simple pathway which children followed as their understanding progressed;
- even where children seemed to be voicing the same conception they differed in how they represented it and what they meant;
- each child had a relatively coherent way of approaching learning and knowledge, linked with the ways they saw themselves as learners and school children.

Overall, the research indicates how we not only need to take account of the ideas children bring to their science learning, but also the wider agenda they bring to activities which we as teachers may introduce.

The research into children's ideas has shown several features which will now be considered separately in order to explore their importance in the teaching and learning of science. The way in which children's ideas can be addressed in teaching is referred to in later sections of this chapter, which include reference to constructivist teaching approaches.

Children's ideas can be very different

Children's ideas are personal constructs which are formed from their experience and the interactions they have with other children and adults. Though children often develop similar ideas about natural phenomena, and there is evidence to show that some ideas are common to children in different parts of the world, they do have different life experiences which influence the way their ideas develop. During the study carried out into children's understanding of forces, a sample of 12 children aged between eight and ten were asked individually what they could tell the teacher about gravity. Even in this small sample of children there was a range of different ideas.

> 'Gravity is something to do with space.'
> 'Gravity keeps you up.'
> 'Gravity is in the air.'
> 'There is no gravity in space or water because you can float.'
> 'Gravity in space makes you float around.'
> 'Gravity is not acting on birds or flying things.'
> 'Gravity stops things floating.'
> 'Gravity makes you come down.'
> 'Gravity keeps you on the ground.'
> 'Gravity pulls things down.'
> 'Gravity is a force pulling things down.'
> 'Gravity pulls all objects down.'

Without further discussion with the children it is difficult to tell what they mean by short statements such as these, but they show a variety of perceptions and misconceptions.

The SPACE project suggested ways of categorising children's ideas.

- *Anthropomorphic views* – For example, 'I think the little caterpillar is scrunched up in that little egg waiting to hatch, but while she is waiting she's planning her life.'
- *Egocentric views* – For example, 'We've got to go to bed. We can't sleep when it is light.'
- *Ideas based on a colloquial use of language* – For example, some children believed that you need to eat carrots to see at night. Everyday language carries an implicit concept that vision is an active process, for instance 'cast our gaze' or 'stare daggers'.
- *Ideas based on limited experience and observations* – For example, some children represent the body as a hollow cavity filled to the neck with blood because, when we cut ourselves, blood comes out.
- *Stylised representations* – For example, children may draw the sun with lines radiating out from it, but this does not mean that they understand that light travels in straight lines.

The SPACE project also found that some children expressed ideas which are more closely related to scientific views than those shown above.

The idiosyncrasy of children's ideas reflects their individual life experiences; it is only to be expected that children's limited experiences result in ideas which differ from the accepted scientific view. From the categories of ideas described above, it can be seen that many children's ideas are very far from the scientific view. A common idea, also prevalent among adults, is that plants obtain 'food' from the soil; the use of the word food is confusing, as plants do obtain nutrients from the soil. What is difficult for children to understand is that plant material is generated mainly from carbon dioxide and water. Incomplete understanding of scientific ideas sometimes prevents children from making distinctions between separate scientific ideas; when children see a solid substance turning into a liquid, for example ice melting or sugar crystals dissolving, they often believe that similar processes are in operation.

For explanations of melting and dissolving, see Chapter 7 of Primary Science: Knowledge and Understanding, Learning Matters, 2011.

PRACTICAL TASK PRACTICAL TASK PRACTICAL TASK

Write down your own explanation of the distinction between melting and dissolving. Consider your explanation and identify what children need to know and understand in order to engage with your explanation.

A further feature of children's ideas is that they can be very specific to particular contexts. A study exploring children's understanding of balanced forces demonstrated the way in which children's explanations of the same underlying principle can be different in different contexts. In the study, children were presented with two different situations involving balanced forces, a box suspended from a piece of elastic and a box supported on a bendy bridge. Children were often inconsistent in the way they identified downward and upward forces acting on the box, though the same principle applied in each case (the upward force of the distorted material balancing the downward force of gravity). Children tend to see different situations as instances of different phenomena, so much so that they may switch from one explanation to a contradictory one.

The internet is a great source of illustrations of concepts such as the balanced forces, discussed above. Use a reliable search engine to find these and then examine them carefully to find the examples that suit the needs of your children and that best fit in with your planning. Ask the science subject leader and your class teacher colleagues what software is already available in school and try to gain access to the school's virtual learning environment (VLE).

RESEARCH SUMMARY RESEARCH SUMMARY RESEARCH SUMMARY **RESEARCH SUMMARY**

Venville (2004) carried out a study with a Year 1 class in an inner-London primary school over five weeks when the children were studying living and non-living things.

Interviewer: Is a cat alive?
Anna: Yes.
Interviewer: How do you know a cat's alive?
Anna: Because I've got a cat and cats have got hearts and they can walk and move. They've got feet and they've got a body and face.
[Interviewer shows Anna a picture of a house.]
Interviewer: Is a house alive?
Anna: No.
Interviewer: Why not?
Anna: Because it doesn't move and it ain't got a heart. Because people live inside it and they can't see the heart inside it.
Interviewer: What about that, what's that? [*Interviewer shows Anna a picture of the sun.*]
Anna: A sun.
Interviewer: And is the sun alive?
Anna: Yes.
Interviewer: Why is the sun alive?
Anna: Because it gives us sun and it can move. It's not sunny today, that means it's gone to another place, and tomorrow, if it's sunny tomorrow, that means it moved over to here.
Interviewer: Oh I see. Has the sun got a heart?
Anna: Yes.

In summary, children have different ideas according to their limited experiences. However, children's ideas make logical sense in terms of these experiences, which often means that the ideas are hard to change.

Children's ideas are hard to change

Because children's ideas make personal sense, they can be very stable and resistant to change. Even when presented with evidence to show that their ideas are not sufficient to explain phenomena, children may ignore the evidence, or interpret it in terms of their own ideas. The following Research Summary demonstrates an example of a child's ideas which were resistant to the teacher's attempts to change them; the extract also raises questions about the strategies a teacher might use to effectively challenge children's ideas, which will be referred to in a later section.

RESEARCH SUMMARY RESEARCH SUMMARY RESEARCH SUMMARY **RESEARCH SUMMARY**

Venville (2004, p. 399) comments as follows on the interview extract cited earlier:

Many young children, such as six-year-old Anna, cannot distinguish between living and nonliving things in a way that adults would consider scientifically acceptable. She believes that the sun is a living thing. Anna understands that a cat is living and that a house is not living. She uses sensible, everyday criteria for these decisions such as the presence or absence of body parts and movement. During Anna's science lessons, the children in the class decided that all living things breathe, eat, drink, move, and have babies. During a discussion, the students in Anna's group agreed that the sun does not need food, grow, or have babies, and the teacher explained that because of these reasons the sun is not living. After the lessons, however, Anna continued to believe that the sun is alive.

Over the five weeks that children were learning about living things, Venville found various types of change in their ideas, but the predominant pattern of learning was the assimilation of facts and information into the children's preferred theory. Children with non-scientific theories of living things were identified as being the least able to benefit from socially constructed, scientific knowledge.

Children – and indeed many adults – find that many scientific ideas are counter-intuitive, in that they seem contrary to everyday experience. A typical example that is difficult for many people to understand is that heavy objects fall through the air at the same rate as lighter objects unless the lighter objects are slowed by air resistance. Because so many objects, like feathers, are slowed by air resistance, people's experience tells them that heavy objects, like hammers, must fall more quickly.

REFLECTIVE TASK

Think of an idea that you have had which you found hard to change to a more scientific view. Write down your idea and an explanation for why you found it so resistant. In a small group, compare and discuss the difficulties that you and your colleagues experienced.

Having described some of the features of children's ideas in this and the previous section, the following sections focus on ways in which children's ideas can be elicited and their misconceptions addressed during teaching.

Eliciting children's ideas and recognising their misconceptions

When considering children's ideas, it is important to understand the distinction between partial understandings and those ideas which are misconceived. Children construct ideas which make sense to them personally, on the basis of their life experiences. Looking back at the ideas children expressed about gravity, one can see why children believe that gravity does not act on birds as they fly – because the birds are not pulled down to earth. Yet how does a child come to say that gravity keeps you up? With ideas such as gravity, which include specialised scientific terms, one can perhaps assume that children have misinterpreted what they have heard, or have been given insufficient opportunity to establish the accepted scientific meaning. Other ideas have been constructed on the basis of experience and are hard to change because they involve structures which cannot be seen directly. It is important to be aware that children's ideas form in different ways, from their exposure to language and meaning and from their limited experiences. In helping children to develop a more scientific understanding of any particular idea, teachers need to find ways of recognising children's misconceptions and challenging them.

There are various methods that can be used for finding out about children's ideas which have been adopted in research and are used in teaching. The SPACE project used informal classroom techniques for eliciting children's ideas. The techniques included the following.

- *Using logbooks* – children could make drawings or do writing over a period of time, to provide a record of long-term changes.
- *Structured writing or drawing* – children could write or draw responses to particular questions from the teacher. Sometimes drawings could be annotated to clarify explanations.
- *Completing a picture* – children could be asked to add relevant points to a picture.
- *Individual discussion* – teachers could ask open questions and listen to children talking about their ideas.

These techniques can be used in the classroom for exploring children's ideas in a range of science topics. In addition, the SPACE team used exploratory activities to find out about children's ideas. For example, in the topic of electricity, children were provided with simple electrical materials, i.e. light bulbs, wires and batteries, and then asked to draw the connections that would be needed to light a bulb on a pre-drawn diagram of a bulb and battery. The drawing in Figure 4.1 illustrates an example from a nine-year-old child.

Figure 4.1 A child's drawing of a bulb and battery

PRACTICAL TASK PRACTICAL TASK PRACTICAL TASK PRACTICAL TASK

The SPACE team concluded that this drawing reflects an understanding which sees the battery as a source of power, the light as the consumer and the wire as a necessary link to enable the supply. Discuss this interpretation with a colleague. How might you be sure the child has this understanding?

Finding out about children's ideas requires a learning environment in which children feel free to express their views. Some children require more confidence than others to contribute their ideas, particularly if they feel that the teacher wants the 'right answer'. Work undertaken by Naylor and Keogh (2000) over many years has addressed this issue. Their highly successful concept cartoons were designed to help children talk about 'other' children's ideas. More recently, Naylor and Keogh have launched a major initiative to use large puppets to enhance children's engagement and talk in science. Research into the use of puppets has shown these to be a useful teaching strategy for eliciting children's ideas.

RESEARCH SUMMARY RESEARCH SUMMARY RESEARCH SUMMARY **RESEARCH SUMMARY**

Research and development into the use of concept cartoons (Naylor and Keogh, 2000) has shown these to be highly successful devices for eliciting children's ideas in science. Building on this success, Keogh *et al.* (2006) undertook research into the use of puppets to promote engagement and talk in science. In this project, primary teachers used large hand-held puppets to create characters that would express alternative ideas, so that children would more readily respond and express their own ideas. The research has shown that children are not only keen to tell the puppet what they know, but also listen more attentively when the puppets are used. Children are more motivated to ask or answer questions and join in discussion.

Analysis of classroom talk before and after using puppets has shown that children are found to use more reasoning when they talk in response to the puppets, justifying their ideas about scientific concepts. Moreover, puppets have been shown to engage children who are usually reluctant to contribute. Puppets have now been used successfully across the range of four to eleven years.

The research with puppets has helped to confirm the recent emphasis on the importance of talk and social constructivist ideas in primary education. A useful booklet by Alexander (2004) discusses the issue of talk in UK primary schools, where he raises the question:

> *Do we provide and promote the right kind of talk; and how can we strengthen its power to help children think and learn more effectively than they do?*

The theme of different kinds of talk in science classrooms is also usefully discussed by Asoko and Scott (2006), and by Mercer *et al.* (2004) in their research, which has shown that children using 'exploratory talk' are able to achieve a better understanding of science.

THE BIGGER PICTURE THE BIGGER PICTURE THE BIGGER PICTURE

When you are planning work in science, remember that talk is important in all areas of the curriculum. Think about the work of Keogh *et al.* (2006) (described above) on using puppets to encourage talk in science and consider whether puppets could be used in other subjects and areas of the curriculum. For example, could they be successful in encouraging children to give explanations of their work on difficult concepts in maths, or to encourage discussions of sensitive issues in personal, social and health education?

A constructivist teaching approach

Once children's ideas have been elicited and the teacher has recognised their misconceptions and partial understandings, decisions need to be made about how to challenge these ideas and help children to develop a more scientific understanding. Such decisions are central to teaching approaches based on a constructivist view of learning. To teach constructively, the teacher aims to adopt strategies which challenge children's existing ideas and enable children to construct new knowledge for themselves. Though the research studies cited previously have all been influenced by a constructivist perspective, there are different theories about the ways in which children change their ideas. Harlen (2000) proposed a model of learning which places an emphasis on the importance of process-skills such as hypothesising, predicting and interpreting in changing children's ideas. In its discussion of students' ideas about particles, the CLISP team focused on the need for children to distinguish the difference between observation and theory, and emphasised the importance of theoretical models in developing children's scientific understanding. Reference to these studies is required for a more extensive discussion about how children construct scientific knowledge and understanding.

As part of their research programme, the SPACE team developed intervention strategies for helping teachers to address children's misconceptions and partial understandings.

RESEARCH SUMMARY RESEARCH SUMMARY RESEARCH SUMMARY **RESEARCH SUMMARY**

The SPACE research reports included accounts of how teaching interventions were planned to challenge children's ideas and help them to construct a more scientific understanding. For example, to help children understand the idea that light travels in straight lines, they were given the problem: 'How could you make the light go round every side of the table?' A strong torch, mirrors and modelling clay were provided and the children had to discuss a preliminary solution before attempting the activity. The drawings they produced were used for discussion with the teacher.

The techniques developed in the SPACE project characterise a constructivist teaching approach. These include:

- building on children's ideas through investigation, posing questions for children to consider and providing opportunities to test ideas;
- testing the 'right' idea alongside the children's ideas;
- making imperceptible changes perceptible, for example using evidence such as videos of time-lapse photography of plant growth;
- helping children to generalise from one specific context to others;
- refining children's use of vocabulary.

The potential of drawing is explored in an article by Brooks (2009):

> *Drawing and visualisation can assist young children in their shift from everyday, or spontaneous concepts, to more scientific concepts. When young children are able to create visual representations of their ideas they are then more able to work at a metacognitive level. When children are encouraged to revisit, revise and dialogue through and with their drawing they are able to represent and explore increasingly complex ideas.* (p. 319)

REFLECTIVE TASK

Refer back to Anna's ideas about living things. Discuss with a colleague different ways in which Anna's teacher could help her to change her ideas.

In order to extend their understanding of science concepts, it is useful to teach children that ideas which apply in one context may also apply to different subject matter. For example, the same principle lies behind the evaporation of water in sweating, drying paint, puddles drying up and the water cycle. Analogies can also be used to help make complex scientific principles more comprehensible. The following classroom story provides an example of a teacher using an analogy to help children understand that light enters the eye through the pupil.

IN THE CLASSROOM

Fiona asked the children to draw how they see a book. She then asked them to think about their drawings and to tell the whole class how they thought they see things. Some children said 'eyes' and 'light', which Fiona acknowledged, then one child said 'you need that little black dot in front of your eye'. Fiona asked the children if they knew what this black dot was called. One child gave 'pupil', so Fiona then asked the children if they knew what the pupil was. They could not answer this other than suggest 'to make you see', so Fiona used this idea of the pupil to provide the children with a useful analogy which would help them to understand how light enters the eye. She said:

It's actually a hole. Imagine your eyeballs are ping pong balls. You've got a hole in the front and a cover on it so things don't get in; what you can see there is actually a hole. How many of you have been to the doctor's or been to see a nurse and they've looked into your eyes? The thing is they are actually looking inside the hole at the back of your eyeball. (Adapted from Simon, 2007)

Though analogies are useful, they can have limitations and children may take them too far. A commonly used analogy when teaching about electric current is that the current is like water flowing around a circuit. However, if the circuit is broken, electric current does not leak out in a puddle as water would. The same caution applies to the use of physical models, though these are essential in science teaching. Models are used to represent phenomena which are too large, too small or difficult or impossible to see; a globe can be used to represent the Earth; marbles can be used to show the arrangement of particles in solids, liquids and gases; and a model torso can show the organs of the body. When concrete models are used, teachers need to point out their limitations.

EMBEDDING ICT EMBEDDING ICT EMBEDDING ICT EMBEDDING ICT

ICT can be of great use in demonstrating models and simulations of a range of key scientific concepts, but these are not just to be watched passively. Don't forget to stop and review key teaching points, to question the children to elicit their views and check their understanding, to challenge any misconceptions and to engender high quality discussion of the clip or program that you are using.

Working from children's ideas can be challenging. Once a teacher has elicited children's ideas, decisions have to be made about how to respond to and value these ideas. Also, there are management decisions about how to group children for the purpose of testing out their ideas. Should children with the same ideas be grouped, or is it better for children with different ideas to work together? It may be that disagreement between children enables them to develop their ideas further. There are other issues to be considered relating to the status of knowledge in the classroom. For example, making children's ideas the centre of attention by asking for and accepting them may serve to reinforce them. In addition, a constructivist approach may not result in children reaching a more scientific view, in which case the children may be in the process of guessing the right answer in order to produce it in a test. Teachers need clear objectives for learning within a constructivist approach, and strategies for differentiation which enable children to learn through appropriate discussion and activity.

A SUMMARY OF **KEY POINTS**

> **Children's ideas are constructs which are formed from their experience and social interactions.**

> **They can be very different due to children's individual experiences.**

> **They make logical sense in terms of these experiences.**

> **Children's ideas are strongly held and can be resistant to change.**

> **They are often significantly different from the accepted views of scientists.**

> **A constructivist approach to teaching emphasises elicitation of children's ideas.**

> **It focuses on the identification of misconceptions and partial understandings.**

> **It advocates testing out children's ideas and alternative ideas through investigation.**

M-LEVEL EXTENSION > > > > M-LEVEL EXTENSION > > > >

As you get to know the children who you are working with better, you will start to perceive patterns and commonly held beliefs about the physical world as you have opportunities to explore the children's ideas in more detail. How will you challenge any misconceptions that they may hold without losing their trust or discouraging them from engaging with science? Use the research summaries ideas for further reading and then try out some of the techniques suggested in this chapter and in your own wider reading.

REFERENCES REFERENCES **REFERENCES** REFERENCES REFERENCES

Alexander, R. (2004) *Towards Dialogic Teaching: Rethinking Classroom Talk*. York: Dialogos.

Asoko, H. and Scott, P. (2006) Talk in science classrooms, in Harlen, W. (ed.) *ASE Guide to Primary Science Education*. Hatfield: Association for Science Education.

Brooks, M. (2009) Drawing, visualisation and young children's exploration of 'big ideas'. *International Journal of Science Education*, 31(3), 319–41.

Harlen, W. (2000) *Teaching and Learning and Assessing Science 5–12*. London: Paul Chapman Publishing.

Harlen, W. (2007) The SPACE legacy. *Primary Science Review*, 97, 13–16.

Keogh, B., Naylor, S., Downing, B., Maloney, J. and Simon, S. (2006) PUPPETS bringing stories to life in science. *Primary Science Review*, 92, 26–28.

Mercer, N., Dawes, L., Wegerif, R. and Sams, C. (2004) Reasoning as a scientist: ways of helping children to use language to learn science. *British Educational Research Journal*, 30(3), 359–77.

Naylor, S. and Keogh, B. (2000) *Concept Cartoons in Science Education*. Millgate House Publishers.

Simon, S. (2007) Children's ideas in science, in Sharp, J., Peacock, G., Johnsey, R., Simon, S. and Smith, R. (eds), *Primary Science: Teaching Theory and Practice*. Exeter: Learning Matters.

FURTHER READING FURTHER READING **FURTHER READING** FURTHER READING

Dawson, V. and Venville, G. (eds) (2007) *The Art of Teaching Primary Science*. Crows Nest, NSW: Allen and Unwin. Chapter 2 gives a readable, up-to-date account of learning which includes constructivism, social constructivism and other aspects such as beliefs and feelings. Chapter 3 gives guidance on how to identify children's science ideas and to track their developing concepts, and indicates appropriate teaching strategies.

Harlen, W. (2000) *Teaching and Learning and Assessing Science 5–12*. London: Paul Chapman Publishing. This is the third edition of Wynne Harlen's comprehensive overview of many issues discussed in this chapter. It is just one of many books that she has written which provide valuable advice for teachers, building on insights from extensive research such as the SPACE project.

Harlen, W., Macro, C., Teed, K. and Schilling, M. (2003) *Making Progress in Primary Science*. London: RoutledgeFalmer. This contains activities that teachers and student teachers can use to improve their practice, such as module 8, which uses examples of children's concept maps, drawing and writing.

Sharp, J. (ed.) (2004) *Developing Primary Science*. Exeter: Learning Matters. The sections on Knowledge and understanding, Life processes and living things, and Physical processes each provide research evidence and many more examples of issues introduced in this chapter.

Skamp, K. (ed.) (2004) *Teaching Science Constructively*, 2nd edn. London: Harcourt Brace. This contains a collection of articles focusing on ways in which different science topics can be taught using a constructivist approach.

Tytler, R. and Peterson, S. (2005) A longitudinal study of children's developing knowledge and reasoning in science. *Research in Science Education*, 35, 63–98.

Venville, G. (2004) Young children learning about living things: A case study of conceptual change from ontological and social perspectives. *Journal of Research in Science Teaching*, 41(5), 449–80.

5
Science in the Foundation Stage

The guidance accompanying these standards clarifies these requirements and you will find it helpful to read through the appropriate section of this guidance for further support.

Introduction

By Early Years in England, we mean from the child's third birthday to the end of the Reception year (a period of two to three years), which covers what is known as the Foundation Stage. The Early Years Foundation Stage (EYFS) (DCSF, 2008a, b) sets the expectations for learning, development and care for children from birth until the end of the August following the child's fifth birthday (for most children this is the end of the Reception year in primary school). The Practice Guidance for the EYFS splits the ages: birth–11 months, 8–20 months, 16–26 months, 22–36 months, 30–50 months and 40–60+ months (which contains the objectives for the Foundation Stage – the Early Learning Goals). We will follow accepted practice here when we refer to practitioners rather than teachers as we include any adult who works in 'out-of-home' provision for Early Years children (i.e. child minders, playgroups, pre-schools, school Nursery and Reception classes, private day nurseries, family centres and independent schools).

As mentioned at the start of this book, at the time of writing this edition, there are possible changes to the Foundation Stage. The government that took office in May 2010 was planning a reform of Early Years policy that will affect the EYFS. Any changes to the EYFS Framework will be likely to have a knock-on effect on the assessment of children in the Early Years, and there has already been much debate about this. In any transitional period, you will need to understand about what curriculum and assessment requirements were in place before the new arrangements and teachers you work with may retain in their practice elements of the earlier ways of working. Therefore, this chapter includes references to the Early Learning Goals for children in the EYFS. Also because of these likely changes, we have focused on giving you insights into the development of the theory behind science teaching in the Early Years.

The development of science in the Early Years curriculum

A number of principles lie behind what we call Early Years education; these relate to its active, experiential nature and the fact that it promotes play and an increasing willingness on the part of youngsters to participate, trial and explore, alone and in social groups. These principles include:

- a unique child;
- positive relationships;
- enabling environments;
- learning and development.

(DCSF, 2008a, b)

The EYFS offers six areas of compulsory experience:

- personal, social and emotional development;
- communication, language and literacy;
- problem-solving, reasoning and numeracy;
- knowledge and understanding of the world;
- physical development;
- creative development.

(DCSF, 2008a, b)

There is a great tradition of science in Early Years education in which children have been encouraged to take an interest in the natural world, to explore it and to engage with appropriate questions. Almost two hundred years ago, Froebel (1782–1852) recommended that structured play was essential for children's cognitive and social development but that this should be enhanced by access to a wide variety of experiences to increase children's understanding of both themselves and the world around them (Bruce *et al.*, 2004). More recently, Montessori (1870–1952) suggested a different child-centred approach where learning is encouraged by having a planned environment which has been tailored to the needs of young children and where there is no time limit on exploration (Pound, 2005).

RESEARCH SUMMARY RESEARCH SUMMARY RESEARCH SUMMARY **RESEARCH SUMMARY**

Research into the development of the brain indicates that it becomes 'wired' between the ages of birth and six years. In other words, the brain develops the neurological connections that it needs to make sense of experiences (Greenfield, 2000). This research suggests that the brain's ability to process information to aid learning is dependent on its physical maturation. Therefore, in any learning situation, it is essential to consider the children's ability to process information. It is our view that science can make a considerable contribution to these processes.

In the EYFS, science is covered under 'Knowledge and understanding of the world' as exploration and investigation, which is *about how children investigate objects and materials and their properties, learn about change and patterns, similarities and differences, and question how and why things work* (DCSF, 2008a).

IN THE CLASSROOM

Mumtaz, a trainee teacher, was asked to share the leadership of a topic on sowing and growing during her Nursery placement in the summer term. She planned activities which linked to communication and number as well as investigation and exploration. In the first week, children were asked to sow seeds and observe others which were germinating. Later on they played in a garden area, counted and measured seedlings, and talked about which seedlings were the tallest, greenest, etc.

'Exploration and investigation' is not taught in isolation from the other topics contained within 'Knowledge and understanding of the world' or indeed any of the other five areas of learning. Connections can be readily made between science, literacy and numeracy and all of these can be developed very effectively alongside one another. It is beneficial, however, that links are utilised and made clear between all the areas of learning, as in the example below.

Consider the example of a group of Early Years children observing and handling snails. Within the six areas of learning below, identify opportunities presented for learning and how these link to investigation and exploration.

Knowledge and understanding of the world	Problem-solving, reasoning and numeracy
Physical development	Communication, language and literacy
Personal, social and emotional development	Creative development

While not always referred to as science, the experience and learning of young children is recognisable to the science educator. For example:

- children bouncing balls on different surfaces to see which is the best surface;
- children using magnifying glasses to observe invertebrates collected (e.g. woodlice from a log pile in the school grounds);
- children using torches to explore reflection and whether some materials are reflective in the dark.

At the end of the Foundation Stage in school the expectations in England are that children whose achievement is average or better will have achieved the Early Learning Goals. For exploration and investigation these are as follows.

- Investigate objects and materials by using their senses as appropriate.
- Find out about, and identify, some features of living things, objects and events they observe.
- Look closely at similarities, differences, patterns and change.
- Ask questions about why things happen and how things work. (DCSF, 2008b)

For each of the four Early Learning Goals above, identify activities that would support children's learning of the topic of growth.

Growth is a topic that many children find difficult to understand and there are many misconceptions both about plant growth and growth in humans. As the result of a survey of teachers' views Pine *et al.* (2001) reported some interesting children's misconceptions that were identified by the teachers.

Misconceptions associated with plants.

- Larger plants are healthier plants.
- Plants will die if they are not kept on a windowsill.
- Seeds come from a packet rather than a mature plant.
- Seeds contain a baby plant.

Misconceptions associated with humans.

- A person's height depends on their age.
- All men are taller than women.
- A person becomes bigger on their birthday.

We all hold misconceptions of various kinds. When practitioners encounter them they should consider ways to challenge them. This might, for example, involve new experiences for the children where they can see if their idea applies in all cases.

Science process and content

Teachers teaching science in the primary and secondary age groups are constantly reminded to emphasise science exploration and investigation. In the Early Years there is no other science; science here is all about exploring and investigating.

Exploration and investigation in the EYFS (DCSF, 2008a) is focused on young children observing, experiencing, finding out, comparing and testing because these are the ways that children learn, through activities that engage all their senses.

ELGs	Science skills
Investigate objects and materials by using their senses as appropriate	Enquire by handling, trialling, testing and observing Say, draw or act out what we have found Fair testing, predicting, speculating
Find out about, and identify, some features of living things, objects and events they observe	Find out by observing, talking about, comparing Develop and strengthen curiosity
Look closely at similarities, differences, patterns and change	Observe, compare, classify, record findings, e.g. draw, model
Ask questions about why things happen and how things work	Ask questions, predict, analyse, interpret Applying ideas, suggest explanations, draw conclusions Improvising, modelling

Table 5.1 Early Learning Goals for exploration and investigation and related science skills (DCSF, 2008b)

Once, an exasperated Early Years teacher planning science with a Nursery class was heard to say *we need to really focus this topic on something manageable.* After some head-scratching, she announced *we'll do the whole wide world*! This in some way is the dilemma of teaching young children. They have everything to learn, they want to know about everything and yet as practitioners we have to construct a learning programme. There is no list of content but typically practitioners will design learning experiences involving toys, clothing, plants, animals, paints, torches, construction materials, making sounds, etc., achieving a balance between biology, materials and physical processes, all delivered through active exploration and investigation. These will be presented in various forms of free or structured play or more formal activities with other children and adults in contexts which are meaningful to children. Building a den in an outdoor or indoor area will require choices about materials: practitioners might offer a range of sheet materials for a roof and children might be asked to try different ones and to talk about the advantages of, say, fabric over paper, or a nylon material over hessian. Options might then be considered for windows and even curtains.

The content of science in the Early Years should draw from the materials, experiences and phenomena with which children are familiar (e.g. water play, toys, familiar animals and construction play in contexts which are meaningful, e.g. my home, my family, weather, children's stories, our school). This should then extend to less familiar examples, materials, experiences and phenomena (e.g. a nature reserve, unusual fruit, making popcorn in a microwave oven, investigating ice balloons, following a sound trail, investigating shadows and exploring bubbles).

It will be important that there is a balance of experience between biological science, materials science and physical aspects of science.

Science aspect	Possible initial exploration/ investigation	Possible follow-on experience
Life processes and living things	Exploring the shape and colour of leaves	Comparing growth in seedlings of different plants
	Investigating the growth of seedlings in class	Growing seeds in different conditions
	Observing snails with magnifying glasses	Observational drawings of snails
Materials	Observation of melting chocolate	Making chocolate crispies
	Investigating which materials are waterproof	Designing an outfit for teddy
	Exploring ice balloons	Recovering objects from inside the ice balloon
Physical processes	Listening to the sound of different mechanical and electrical toys	Investigating the effect on a sound by wrapping a ticking clock in different materials
	Observing our faces in different mirrors	Taping mirrors together to investigate a kaleidoscope effect

Table 5.2 Suggested exploration and investigation activities

THE BIGGER PICTURE THE BIGGER PICTURE   THE BIGGER PICTURE

When you are planning topics in the Early Years, think about planning across all of the areas of learning rather than just for Knowledge and Understanding of the World. Taking some of the suggestions from Table 5.2 above, think about how these might fit into a wider topic or unit of work. For example, observing faces in a mirror might fit in with the theme of 'Ourselves' and investigating which materials are waterproof might form part of a half-term's focus on 'Clothes'. Investigating the growth of seedlings and observing melting chocolate might both be included in a cross-area plan based on 'Food and farming'.

Exploration and investigation

Exploration can be a less formal approach to science. While it can be systematic, it can also look very much like play because children are often experiencing something they have had limited or no contact with before. They often benefit from exploring alongside or with others, sharing their observations and prompted by the actions, thoughts and questions of other children and adults. Exploration involves a physical element, observing, handling and experiencing. Importantly, it includes a cognitive element where observation involves senses and the mind, making comparisons with previous experiences, spotting differences and similarities and verbalising what is observed. Investigation in the Early Years tends to be more formal, and is often based on answering a question, usually posed by the practitioner but sometimes in response to a question or issue raised by a child. It is likely to involve finding out about why things happen or how things work, and is likely to give children opportunities to observe changes and identify patterns.

RESEARCH SUMMARY RESEARCH SUMMARY RESEARCH SUMMARY **RESEARCH SUMMARY**

There have been several seminal theories on the importance of play that underpin current approaches in the Early Years. Piaget (1954) viewed play as a way for children to develop and refine their ideas before they were capable of thinking them through in the abstract. He therefore emphasised the need for children to explore and investigate for themselves – the 'lone scientist'. Vygotsky (1978) also considered play to be important for children's cognitive development but he emphasised the social and cultural aspects. In particular, he believed that play allowed children to think in more complex ways than they could in their everyday lives and to use their imaginations to take part in impossible adventures.

The EYFS (DCSF, 2008a) requires that all children have the opportunity to play in safe and secure environments.

REFLECTIVE TASK

It may be that play is undervalued in other phases of education, but in Early Years settings it is celebrated and exploited by the practitioners who understand its value.

> *The child's exploration of the world is the springboard from which the next leap is taken, that of more systematic enquiry. Systematic enquiry may be described as 'science investigations' in later years. Nevertheless, the first step in any scientific enquiry is exploration, or 'play'.* (de Boo, 2000, p.1)

Make two lists. One should contain all the activities you think of as 'play' in science but try to describe what the children do as fully as possible. The other list should contain all the activities you think of as 'work' and again try to explain the activity in detail. Check for words that you use frequently in each list and then try to write definitions of 'play' and 'work'.

IN THE CLASSROOM

Mrs Burgess, the Nursery teacher, had noticed the children looking at their reflections in the shiny metal of a spare serving spoon at lunchtime. She decided to let the children explore their reflections in a more structured way later that afternoon by putting out various reflective materials including plastic convex and concave mirrors.

David: I'm big. I can see ... big just like me.
David turns the spoon over.
Upside down. I can see Josh.

Mrs Burgess: Josh is upside down as well?

David: Yes. I see him upside down.

Josh: No!

So that the children could record what they had observed, Mrs Burgess had supplied coloured pencils and sheets of paper with two circles drawn on each one to represent the outline of faces as seen in the curved mirrors.

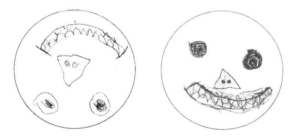

Figure 5.1 David's drawings of himself in the concave and the convex mirrors respectively

Figure 5.2 Josh's drawings of himself in the convex and the concave mirrors respectively

When Mrs Burgess asked Josh why he had included his feet, he held up his mirror, tilted it towards the ground and said, *Because I sees them*. She did attempt to explain that if he held the mirror in this way he could no longer see his face but he was adamant that his feet should be included in his drawing.

> **THE BIGGER PICTURE**   THE BIGGER PICTURE   **THE BIGGER PICTURE**
>
> When you are planning science work in the Early Years, remember that sometimes unexpected opportunities will arise, as in the example above, to do some investigative activity based on a spontaneous event or comment of one or more of the children. Be prepared to 'seize the moment' and to respond while the high levels of engagement are there and then to build it in to your more formally planned lessons as and when appropriate. Never be too rigid in adhering to your original ideas – you may miss a really valuable and exciting learning opportunity!

Children's understanding of science

Young children will already have learnt much about the world but will sometimes find that what we call the 'scientific view' challenges their ideas. In these cases it will be the role of adults to challenge the children's 'alternative' viewpoints. An example of this is often seen when practitioners discuss whether human beings are animals. Very many young children will challenge this as they have a view of animals based on pets, wildlife, zoos and farms, even if these have only been observed on TV.

Young children may have a very patchy knowledge and understanding of the world based on their life experiences to date. They are unlikely to use words such as 'material' or 'energy' in a strictly scientific way.

> **IN THE CLASSROOM**
> As part of a topic on caring for others, Dee asked her children to observe butterfly eggs which the children would observe developing into caterpillars, pupae and then butterflies. One child was very excited by the eggs and what would happen, but referred to the eggs as shells. Dee then brought into school other eggs (hen and duck) and a collection of sea shells so that the children could handle them, describe them and consider their purpose.

Organisation

When organising what the children will do each day in an Early Years setting, you should consider planning for both adult-led activities where it is you, the adult, who guides and directs the children's activities and also child-initiated activities where it is the children who choose what they will do.

It is essential that you provide flexible resources that the children can make use of in many different ways in order to facilitate both their play and exploration. You will therefore need to provide an assortment of materials that can be used for different purposes and objects that work in different ways. In order that the children can record their observations and/or discoveries, you should also provide writing and drawing materials, resources for making models and even a camera.

EMBEDDING ICT

ICT is an important tool in science because it motivates children to learn by engaging their interest, by encouraging them to share their ideas, and by being fun! Digital cameras, digital recorders, tape recorders, camcorders and webcams are continually becoming cheaper to buy. Remember, you do not need the most sophisticated equipment; basic models are generally far easier for young children to use. There is also software available that makes it possible for young children to transform digital photos into a slide show with captions and audio.

It should be remembered that the outdoor classroom is equally important and ideally you should plan for play and learning opportunities that flow seamlessly between the inside and outside environments. This organisation will enable the children to build on their ideas without unnecessary interruption.

Many spontaneous events, such as a fall of snow, also provide the opportunity for children to explore and investigate. These opportunities should never be ignored because the unusual nature of the event makes it memorable and consequently the children will be highly motivated to learn.

Areas are often established as contexts for learning. These might include a role-play workplace such as a shop, doctor's surgery, or a play context linked to a story such as *Mrs Wobble the Waitress* or *Dear Zoo*. A science area might contain materials to handle and explore, such as reflective materials, or equipment for exploring and investigation, such as lenses, colour paddles and a kaleidoscope.

IN THE OUTDOOR CLASSROOM

The children rushed over to Louise and excitedly told her about something furry that they had found crawling across the flowers in the garden. Louise followed them back to the place where they had seen this furry visitor. It was a caterpillar. They all watched it for a few moments. Louise asked them what they could see and about the features of the caterpillar's body. When they returned to the classroom, Louise asked them to draw what they had seen. Later, she taught them the rhyme about Arabella Miller who found a hairy caterpillar.

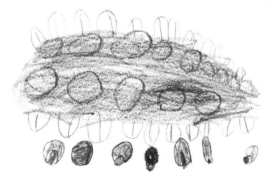

Figure 5.3 Observational drawing of a caterpillar crawling across flowers

The role of adults

Practitioners working in the Foundation Stage have to be extremely flexible so that at times they can take charge, direct and instruct, and at other times allow children considerable freedom to interact, create and explore. Adults plan and organise learning experiences and spaces indoors and outdoors to provide a full breadth of experience and ensure that children have the opportunity to develop physically, emotionally, socially and cognitively. Science provides excellent opportunities for all of these. Practitioners might:

- introduce a context for exploration and/or investigation;
- introduce a new material, piece of equipment or experience;
- persuade children to slow down, look and listen;
- instruct children developing a new skill (e.g. to use a magnifying glass);
- elicit children's present understanding and previous experience;
- support children in articulating their observations and thoughts;
- encourage children to reflect on their discoveries through drawing, painting, modelling and constructing.

The following aide-memoire is intended to act as a prompt for adults encouraging young children to behave scientifically. It can be helpful for all staff, students and trainees, and any parent helpers, whether in school or on an educational visit.

Foundation Stage aide-memoire for staff: Scientific exploration and investigation

Make sure you know the objectives of the session. ☐

Link the science investigation to the real world and their lives. ☐

Look for opportunities for the active involvement of everyone. ☐

If materials, etc., are new to the children, give them an opportunity or activity to familiarise themselves and talk about features. ☐

State simply and clearly what we are going to do (write it down). ☐

Ask for ideas about what will happen (predictions) and record these to come back to later on. ☐

Show, name and discuss any equipment and materials (point out possible danger/ask them to suggest how we can keep safe). ☐

Ask children for suggestions about what you will do and what equipment will be needed. ☐

Ask children for ideas about the order of actions. ☐

Recap on the question, the prediction and the procedure (allow more handling of equipment). ☐

Start the test involving the entire group if possible. ☐

Ask open questions about what is happening: *Tell me what you're doing. Tell us what you can see. How will you do this? What will happen next?*	☐
Encourage speaking and listening.	☐
Perhaps stop the activity to demonstrate progress/a teaching point.	☐
At the end, get children to say what happened/what they found out.	☐
Go back to the predictions and ask if they were helpful.	☐
Conclude with children being asked to say what was done and what was found out.	☐
Finally, link this to real life and their own lives.	☐

The importance of speaking and listening

To all scientists, language is important. In science education language and communication are important. In the Early Years it is impossible to overemphasise the contribution of language in science education.

Children should be encouraged to share their thoughts and ideas about the things they observe in science-related activity. An increasing science-related vocabulary will further assist learning in science-related areas but also in the other areas of learning. With this in mind, opportunities to engage young explorers and investigators in speaking and listening should be maximised. Science can provide wonderful and amazing experiences, so most children should become eager to participate and talk.

Food tasting is an enjoyable experience for most children and it can offer excellent opportunities for trying out their descriptive abilities and for learning new vocabulary, providing that you take into account children's allergies and any religious dietary needs. Each child can be given the opportunity to taste the food and should be encouraged to comment on how it tastes; making a face will not suffice. As a result of the activity, including the opportunity to discuss their likes and dislikes, the children will learn the names of foods that they do not normally encounter in their everyday lives. In addition to promoting children's speaking, the activity should also improve the listening skills of the children waiting for their turn to taste the food.

Fore more on Health and Safety in this activity, see Chapter 11.

REFLECTIVE TASK

The Rose Report (2006) emphasised the importance of speaking, listening, reading and writing in the EYFS. These communication skills are considered to be of prime importance to beginning readers and therefore should be considered in any science planning.

Think about ways in which these four skills can be included in your science activities with children.

Safe working

See Chapter 11 for more on Health and Safety.

Safety is, of course, the responsibility of all practitioners and was at the heart of *Every Child Matters* (DfES, 2003). In science, children are encouraged to explore. Practitioners must ensure children are safe but also that they are educated to be prepared to take on more and more personal responsibility. Children must learn to use their senses and handle things, which is a good opportunity to ask them to deal with risk. It is essential that they are safe at all times but that they are given opportunities to consider what they are doing and are encouraged to consider the risks involved, not simply the threats to their own safety but also the safety of others.

Before deciding on the foods for any tasting activity or for cooking generally, you need to find out if any of the children have food allergies or if there are any cultural considerations to be observed. Parents/carers will have notified the key person responsible for their child in the setting and this information should therefore be readily available.

IN THE CLASSROOM

Jodie, an NQT working with a Nursery class on a topic about celebrations, wanted children to observe a range of candles burning. After consulting colleagues and the booklet *Be Safe!* (ASE, 2001), she organised a series of metal trays containing sand into which she placed a range of stable candles. She ensured that she had a bucket of water close at hand and that she could work with small groups of six to eight children.

When working with the children, she explained what they were going to do and asked children to talk about celebrations and times when they burned candles. She talked to them about the danger of fire but that because they were with an adult they could be safe as long as they were sensible. She elicited from them sensible rules for proceeding.

PRACTICAL TASK   PRACTICAL TASK   PRACTICAL TASK   PRACTICAL TASK

Consider safety issues and how you might deal with them if a class were going to collect leaves on the school site (it is illegal to collect any plant materials outside cultivated gardens). When you have done this, consider whether young children could think about danger to themselves and others.

Resources

Resourcing for 'Exploration and investigation' does not have to be expensive. Many resources can be made by you or provided by parents and carers. If you hang a 'wish list' on the outside of the classroom door, you will be surprised at how much help you will receive. You should always consider health and safety and as a teacher you are acting *in loco parentis*, that is, as a responsible parent would, so you should not ask children to handle glass or other dangerous materials.

Here are some suggestions for equipment and resources that can be used in scientific explorations and investigations:

Specialised equipment: torches, batteries, magnifiers, feely bags/boxes, selections of different materials, magnets, magnetic train/trucks, plants, seeds, plant pots, compost/soil, live animals (goldfish, snails), pictures of animals and minibeasts.

Other resources that can be used for science: construction equipment for both indoors and outdoors, e.g. wooden blocks, magnetic blocks, construction sets.

Water play: plastic fish and boats, wheels, jugs, containers, sieves, sponges, pipes, funnels, filters, real shells, items for floating and sinking, watering cans, guttering, buckets.

Sand play: cups and saucers, sieves, jugs, containers, plastic minibeasts and animals, plastic spades, rakes, buckets.

Musical instruments: include multicultural instruments.

Sorting: buttons, fir cones, shells, stones, cotton reels.

Measuring: tape measures, sand timers, scales/balances.

Mark-making materials for recording findings: paper, pencils, crayons, chalks, brushes and paint, collage materials, glue, large playground chalks for marking shadows and around puddles outside.

Books of all types: either with a science theme or that can be used as a 'starter' for an activity.

Role play with a science theme:
- forces – space station, garage, aeroplane;
- plants – garden centre, florist;
- habitats – the jungle, under the sea, the zoo;
- humans – hospital, dentist, fruit and vegetable shop (healthy eating);
- light – cave, shadow puppet theatre;
- materials – shop selling cooking ingredients, builders' merchant.

EMBEDDING ICT EMBEDDING ICT EMBEDDING ICT EMBEDDING ICT

A webcam might make it possible for children to see animals that are difficult to see under normal conditions, for example foxes looking for scraps of food in the evening. Digital photographs displayed on a computer screen allow much closer scrutiny for similarities, differences, pattern and change than could be achieved normally. Audio equipment can be used to capture children's thoughts and ideas. A collection of digital images collected during a hunt for living and non-living things in the school grounds can be sorted and resorted by the children during the following weeks or can be displayed on a wall as a reminder of the event.

A SUMMARY OF **KEY POINTS**

> **Science makes a considerable contribution to the Early Years curriculum.**

> **Young children benefit from a range of experiences, including scientific exploration and investigation.**

> **They come to school with developing ideas about the world.**

> **Science needs to be meaningful and practical.**

> **Young children need time to follow their own interests (child-initiated activities).**

> **Speaking and listening form an important part of every science activity.**

> **Remember that the outdoor classroom contains many valuable resources for young explorers and investigators.**

> **Science needs to be a stimulating learning experience within a safe environment.**

M-LEVEL EXTENSION > > > > M-LEVEL EXTENSION > > > >

Look back at the science explorations and investigations exemplified in this chapter, including in the reflective and practical tasks and 'In the Classroom'. Most have concentrated on growth as part of the Life processes and living things aspect of science. Challenge yourself to develop similar sequences of planning for topics or themes relating to Materials and their properties and Physical processes. Consider whether you can include valid links to the other areas of learning and identify these explicitly in your planning. Include notes on where there will be opportunities to develop children's speaking and listening skills, what resources will be necessary including the use of ICT, and how you will use other available adults to lead/facilitate activities or to support child-led enquiry.

REFERENCES REFERENCES **REFERENCES** REFERENCES REFERENCES

ASE (2001) *Be Safe!* Hatfield: Association of Science Education.

Bruce, T., Findlay, A., Read, J. and Scarborough, M. (2004) *Developing Learning in Early Childhood Education*. London: Paul Chapman.

de Boo, M. (ed.) (2000) Why early-years science, in *Laying the Foundations in the Early Years*, 1–6. Hatfield: Association for Science Education.

DCSF (2008a) *Statutory Framework for the Early Years Foundation Stage.* London: DCSF.

DCSF (2008b) *Practice Guidance for the Early Years Foundation Stage.* London: QCA Publications.

DfES (2003) *Every Child Matters: The Green Paper*. London: DfES.

Greenfield, S. (2000) *The Human Brain*. Guernsey: Guernsey Press.

Piaget, J. (1954) *The Construction of Reality in the Child.* New York: Basic Books.

Pine, K., Messer, D. and St. John, K. (2001) Children's misconceptions in Primary Science: a survey of teachers' views. *Research in Science and Technological Education*, 19(1), 79–96.

Pound, L. (2005) *How Children Learn: From Montessori to Vygotsky*. London: Step Forward Publishing.

Rose, J. (2006) *Independent Review of the Teaching of Early Reading.* London: DfES.

Vygotsky, L. S. (1978) *Mind and Society: The Development of Higher Mental Processes*. Cambridge, MA: Harvard University Press.

FURTHER READING FURTHER READING **FURTHER READING** FURTHER READING

de Boo, M. (2006) Science in the Early Years, in Harlen, W. (ed.) (2006) *ASE Guide to Primary Science Education*. Hatfield: ASE.

Drake, J. (2001) *Planning Children's Play and Learning in the Foundation Stage.* London: David Fulton.

Howe, A. and Davies, D. (2003) *Teaching Science, Design and Technology in the Early Years.* London: David Fulton.

Oliver, A. (2006) *Creative Teaching: Science in the Early Years Classroom.* London: David Fulton.

Stephen, C. (2006) Early years education: Perspectives from a review of the international literature. *Practical Research for Education*, 36, 65–73.

6
Teaching strategies

Introduction

Teachers gain a wide repertoire of approaches and strategies for delivering the curriculum throughout their careers. Varying the approach to teaching is crucial to motivating children to learn and to a teacher's own professional development. This chapter looks at a range of strategies which might be employed to teach science in the primary classroom through a range of practical examples.

Guiding principles for choosing appropriate teaching strategies

- Whatever happens in the classroom, children should become better and better at thinking and behaving scientifically and understanding scientific ideas.

It is possible for the majority of children in a classroom to be happily carrying out science activities for an hour or so but to have learnt nothing new by the end. Teachers need to use their knowledge of the children, based on previous assessments, to plan work which takes them another step on. The work should be challenging, stimulating and demanding. Children should be motivated to learn rather than merely occupied by the experiences provided. When it comes to developing children's procedural skills, these need to be practised in a meaningful context which provides a new challenge at the same time as reinforcing skills already acquired.

- Teachers should gain some insight into the ideas that children hold in a science topic before extending this knowledge.

The National Curriculum for primary science was apparently written in such a way that assumed that all children in a class will have reached a certain level of understanding by a certain time and will all be ready to move on to the next stage together. Research has shown this not to be the case and that children learn by restructuring the ideas that they currently hold only when new information is sufficiently plausible and intelligible. The elicitation of children's ideas before a science topic begins enables the teacher to plan work which is more appropriate for groups of children.

For more on children's ideas, see Chapter 4.

- Teachers should provide a range of experiences which will lay the foundations for helping children understand the big science ideas later in life.

Primary teachers need not think that they should always be able to explain why things are so to children. Often, it is more appropriate to enable children to experience and observe a certain phenomenon without necessarily understanding all aspects of it. If teachers, then, have an idea where these experiences are leading, they will be able to lay suitable foundations for future learning. For instance, children might observe and record the similarities and differences between some liquids and solids so that one day they will be able to understand the particulate nature of all matter. Knowledge of where the children are heading will influence the kind of questions the primary teacher might ask.

- Practical activities should be used when appropriate, but there are instances when alternative strategies should be used.

Children who learn about what is inside the human body or about the Solar System are unlikely to do this completely through practical activities. Secondary sources of information such as pictures, video or the use of models are more likely to be used as well as analogy. Often, a profitable combination of practical activity, together with other strategies such as reading, drawing or discussion, will achieve the teacher's objectives.

EMBEDDING ICT EMBEDDING ICT EMBEDDING ICT EMBEDDING ICT

ICT can provide valuable examples of secondary sources. There are many programs available, both for use with the whole class on the interactive whiteboard (IWB) and for children to use in groups on the class computer, that will make science fun while providing useful models and simulations. Make sure that you know what is available in school and via the local authority's VLE, but also consider the internet. Try some of the following websites:

www.ase.org.uk
www.bbc.co.uk/cbeebies
www.bbc.co.uk/schools/scienceclips
www.britishecologicalsociety.org
www.microsoft.com
www.rsc.org
www.sciencebuddies.org
www.tes.co.uk/green-tv-videos
www.tes.co.uk/primary-forces

- Teachers should be clear about which skills and concepts they want children to learn and bring these to the fore in any lesson.

Children do not soak up skills and knowledge simply by being exposed to a scientific idea through activity. It is important that teachers teach the key ideas directly though asking appropriate questions and directing the children's attention towards the significant parts of a concept.

- Teachers should use language precisely in science so as not to cause confusion. It should be recognised that the language used can contribute to the organisation of understanding in science.

Our everyday use of such words as 'force' and 'plant food' often give rise to misunderstandings in science. Teachers need to be aware of the potential for such confusion and to provide with care good examples of scientific language. The use of the correct linguistic 'labels' can enable children to organise their thoughts on a subject and to convey their understanding to the teacher.

- Teachers should exploit the links between science and other areas of the curriculum.

Cross-curricular links provide meaningful contexts for science, making what appears to be a rather abstract concept more relevant to the children's lives. At the same time, the application of scientific ideas in new contexts, such as a technology lesson or a history lesson, will help to reinforce these ideas. Furthermore, cross-curricular work will enable the teacher to assess children's understanding of science concepts in new contexts. Teachers in schools which have adapted cross-curricular themes should ensure that time is still devoted to teaching science and that links are made to reinforce and enhance learning.

Starting with children's ideas

For more on children's ideas, see Chapter 4.

Good teaching in primary science involves understanding the ideas that children hold on a topic and then planning work which enables them to challenge and reconstruct their ideas. There are a variety of strategies that teachers can use to elicit children's ideas. These include:

For more on these strategies, see Chapter 9.

- observing children investigating;
- discussion with children (individuals, groups or whole class);
- concept mapping;
- structured writing or making annotated drawings;
- creating a poster;
- using a floor book to record ideas in a group;
- discussion cartoons;
- completing a picture or drawing a comic strip to explain a phenomenon;
- giving a short written test.

Teacher exposition

Teacher exposition and explanation has an important part to play in the effective teaching of science. This will often happen more during the introductory or plenary phases of a lesson when the teacher has the attention of the whole class, but of course it may take place while the children work individually or in groups.

Teachers might adopt some of the following strategies when explaining scientific concepts.

- Recap of the children's previous learning in the same science area.
- Focus on the important aspects of the science idea and avoid introducing irrelevant or confusing material.
- Where appropriate, use models, analogies or illustrations to support the teacher explanation.
- Use a physical demonstration to accompany an explanation if appropriate.
- Use a picture or an artefact to base discussion around.
- Pass things round for the children to handle if possible.
- Simplify the science concept so that communication is at the child's level.
- Break down the science idea into smaller, easily understood parts.
- Backtrack and simplify an idea if the children clearly do not understand the first explanation.
- Use question and answer to ascertain the children's general knowledge and understanding.
- Relate the science idea to an event that is familiar to the children.
- Keep teacher explanations short and to the point or use a longer discussion if this is more appropriate.
- Accompany the explanation with a true story if this is appropriate.
- Weave the explanation into a fictitious story if this seems appropriate.

Teacher demonstration

A physical demonstration may be used as part of a teacher's exposition. This is often more easily controlled with the children sitting, listening, watching and participating. Teacher demonstration to the whole class should be used where appropriate and should never take the place of child activity where this is possible. There are a number of points to bear in mind when demonstrating science ideas to children.

- Ensure that all the children can see what you are doing. Try to demonstrate with large pieces of equipment.
- Consider moving your position around the classroom as you demonstrate.
- Make the demonstration short and snappy. (The children will be itching to have a go too.)
- Involve the children with practical tasks and questions.
- Reinforce the main points of the demonstration by following up with a simple child activity or perhaps some form of recording of ideas.
- If possible, after the demonstration, arrange for the children to gain first-hand experience of the equipment you have used, in small groups or pairs.

IN THE CLASSROOM

Mrs Smith wanted to demonstrate to her Year 2 class how the human skeleton enables the body to move through the use of muscles and joints. She had arranged to borrow a large, full-size, plastic skeleton and had it hanging in the classroom as the children came in from playtime. The children assembled on the carpet as the skeleton loomed above them.

Mrs Smith focused on two key ideas in her demonstration: (a) the fact that joints are necessary between sets of bones in order for movement to take place, and (b) the way in which a muscle acts on a bone to pull it into place. All the children were asked to feel the bones in their own bodies as they looked at them on the model and some were asked to pin the names of some of the bones onto the model.

Movement of parts of the body was demonstrated by Mrs Smith and some of the children. The skeleton was made to scratch its head and kick a football to see how many bones had to move. A large elastic band was fixed to the forearm to demonstrate how a muscle might pull it upwards and another showed how another muscle would pull it back.

The demonstration was followed by an activity in which the children made a working model of an arm with elastic bands for the biceps and triceps muscles while small groups of children took a closer look at the model skeleton itself.

Using practical activities in primary science

Primary children enjoy learning through practical activity. They will often claim that 'we haven't been doing any work today' when most of their work has been of a practical nature. This is probably a sad reflection on the fact that a lot of school work is centred on 'sitting and writing'. Science teaching in the primary school lends itself to practical activity so teachers should take advantage of this fact. Practical activity in science means that children have some kind of hands-on experience. For instance, children may be sorting and classifying a collection, making measurements or arranging equipment for an investigation.

Practical work can be planned and organised in different ways. Teachers need to be aware of the variety of approaches which are possible and that there are times when alternative teaching strategies will be more appropriate. Medium-term science planning should ensure a balance of practical work and other ways of learning. Practical work might include:

- a guided **illustrative activity** to teach a concept;
- an **investigative activity**, including planning and carrying out scientific enquiry;
- the **observation** of a particular phenomenon;
- an **activity to develop a particular scientific skill** (including the use of equipment).

These four categories of practical work are not exclusive. For instance, observation activities may form part of a broader science investigation. An example of this might be when children need to learn how to use a stopwatch as part of an activity which illustrates how pulse rate changes with exercise.

Here are further examples of these types of practical activity.

- An activity in which children learn that higher pitched notes are made by shorter stretched elastic bands. (*Illustrative activity*)
- Children make a full-size model of a daffodil to learn about the parts of a flower as well as to enhance their observational skills. (*Observation activity*)
- Children plan and carry out an experiment to find which materials are best at blocking sounds from reaching the ear. (*Investigative activity*)
- A short activity in which children 'play' with a collection of magnets to see what they will do. (*Observation activity*)
- Children make drawings of a snail as it moves over a leaf. (*Observation activity*)
- Children measure the force required to move things such as an open door, or a work tray or to lift a wooden block. (*Skills activity*)
- A problem-solving activity in which children design the best parachute in order to learn more about air resistance and the effects of gravity. (*Involves all categories of activity!*)

Organising practical work in science

Teachers need special skills in order to organise practical work for their children. They will need to answer the following questions.

- Should all children in the class do the same practical work at the same time?
- Should all groups do the same practical work at the same time?
- What sort of groups should the children work in – individual, friendship, ability or attainment?
- How will the equipment be distributed, shared, maintained and collected in?
- How much space is needed for the practical work? Does the furniture need rearranging?
- How much of the lesson should be spent on the practical aspect?
- What will the teacher's role be during the practical work?

See Chapter 8 for a consideration of child grouping and organising resources and time.

Teachers will not stop teaching once their introduction is finished and the children are busily employed. Often, they will take the role of facilitator or consultant as they sort out the problems of missing equipment, misunderstood instructions, social grouping and so on. A much more important role is achieved through the use of questions and explanations which provoke thought, provide new ideas and assess learning. The teacher will want to talk to the children and stand back and observe their actions and thoughts when appropriate rather than spend every minute troubleshooting. This can be achieved through thorough planning and organisation and by encouraging the children to become independent learners.

Another important function of the teacher during a practical science lesson is to maintain pace. The teacher will gain a sense of when to stop an activity to draw together some important point and when to allow the children to continue with their individual work. Having a short discussion about one group's achievements so far will often point others in the right direction and send messages about how the teacher wants the whole class to behave.

Which key points will be learnt during an activity and which key questions will promote learning?

For more on asking questions, see Chapter 7.

The answer to this question lies within the teacher's own subject and pedagogical knowledge and understanding of science. The teacher must know where the children have come from and where they are going with the particular science being taught. The key ideas are spelt out in the learning objectives for the lesson and should be small enough to be digested by most children in the class. It should be made clear to the children what they are expected to learn from an activity, rather than leaving this to chance. The preparation of some key questions to ask in the introduction, development or conclusion of the lesson will help.

IN THE CLASSROOM

Mrs Jones' learning objective for a lesson on pushes and pulls with her Year 1 class was written as follows:

> *At the end of the lesson, most children will have identified three different ways of describing movement – getting faster, getting slower and keeping the same speed.*

The children were given a range of cartoon pictures showing moving objects such as toy cars and model swings. They worked together to reproduce what was in the pictures with cars and swings which had been arranged on the table. They were encouraged to talk about the ways that these moved.

Mrs Jones prepared some key questions/challenges for the lesson, which she shared with her classroom assistant and parent helper:

- Is the toy car moving?
- Can you make it move faster?
- Is the toy car moving fastest at the top of the slope or at the bottom?
- Can you describe how the movement of the swing is changing?

How will the children be given instructions during an activity?
Teachers will need to consider a range of strategies for giving instructions to children as they carry out practical activities. The obvious strategy of explaining what to do in the introduction to the lesson can have its limitations as many teachers have found. Different ways of conveying instructions are as follows.

1. Using regular verbal instructions as the activity proceeds. This, however, assumes the children will all be working at the same pace.
2. Writing and illustrating instructions on the board/interactive whiteboard.
3. Using a commercially produced workcard or poster.
4. Using a teacher-designed or commercially available worksheet. The worksheet might be used:
    - as a list of instructions for carrying out science activity;
    - as a comic strip sequence to explain what to do;
    - to write observation records on;
    - to write answers to questions on;
    - to complete a table/picture/graph;
    - as a planning proforma;
    - as an information text.
5. Briefing a classroom assistant who is assigned to a particular group.

How will the children consolidate their learning during or after an activity? What kind of written records should be expected?
Children might summarise or consolidate what they have learnt from a practical activity in a variety of ways. A discussion of the main points of the lesson, led by the teacher, is a useful way to conclude a lesson. Sometimes groups or individuals can be called on to describe what they have done and learnt. Children might use a variety of recording methods in which they reflect on what they have been doing. A group poster or a word-processed report are just two examples.

See Chapter 7 for more on concluding a lesson.

What sort of behaviour should be expected during an activity?
Teachers will devise their own set of rules for appropriate behaviour in practical science lessons. They will need to decide if it is appropriate for children to talk to others in their group or in the room as a whole. The nature of practical work, often involving co-operation with others, will almost certainly require group or partner discussions but general noise levels may need to be controlled. Teachers must decide if free movement around the room is safe and/or desirable.

See Chapter 8 for more on behaviour.

Secondary sources of information

There are some instances where primary school children may not learn by first-hand experience. Science activities may often involve them in reading or working at a computer. Books, videos and computer or internet sources, as well as listening to and watching their teacher, provide children with secondary sources of information. Teachers should plan to integrate these secondary sources of information in lessons to provide a balance of learning experiences and learning styles. Secondary sources of information are particularly useful where the science would otherwise be inaccessible or dangerous to undertake, where children's knowledge and understanding can be enhanced and extended appropriately, or simply where they provide the most effective way of teaching. Quite often children can consolidate learning from a secondary source by carrying out a practical activity.

IN THE CLASSROOM

Following an introductory lesson where the class viewed a website providing a tour of the Solar System, Manesh and Sarah had used a CD-ROM to discover that the Earth was just one of the planets which orbit the Sun and that the Moon orbits the Earth. Their teacher encouraged them to make a working model of the Earth–Sun–Moon system so that they could fully understand how each body moved in relation to the others.

The children completed their own card model, which enabled the Earth to rotate around a paper fastener and the Moon to orbit the Earth on a disc of card. They were able to use this model to proudly show their teacher how night and day occurred. After some thought-provoking questions from the teacher, they were surprised to learn that the Moon could be present in the sky during the day as well as at night and wanted to know from their teacher why they could not usually see it during the daytime.

Teachers will want to ensure their children get a balance of practical and other experiences in science, with one strategy being used to complement the other.

The use of role play

Role play in which children act out certain scenarios can provide a useful alternative teaching strategy when appropriate. Teachers might use this strategy to promote debate about science ideas such as the misrepresentation of scientific data or to explore contemporary scientific issues. In contrast, children can be encouraged to imagine they are part of a food chain or perhaps an electron in an electrical circuit and explore the science concepts involved in this way.

IN THE CLASSROOM

Miss Appleton's Year 6 class were involved in a unit on 'How we see things'. She wanted them to explore further the idea of reflection by using role play. In a physical education lesson she had them working in pairs to demonstrate the actions of a reflection in a mirror. The children had observed that a right hand raised produced a corresponding movement in the left hand of the image in the mirror. Now they were able to use this information in their role play. At the same time they were able to simulate the information they had gained about the distance of the subject and image from the mirror.

Back in the classroom an interesting discussion developed about the relation-ship between an actor, a mirror and the camera. Was it true that an actor had to look at the camera when they looked in a mirror on screen, and if so, how were they able to put on make-up and comb their hair 'on camera'?

Presentations and debates

Children might work in groups or as individuals to prepare and present a variety of science ideas. This might include the results and conclusions of an experiment, their ideas on how shadows are formed, or their predictions as to which food their collection of snails will prefer. Children can use a variety of multimedia devices to enhance their presentation. Posters, a flipchart or an overhead projector might be employed as well as an IWB. Debates or discussions about the moral and ethical dimension of science have a value in promoting both scientific literacy and an appreciation of a wide range of issues.

THE BIGGER PICTURE   THE BIGGER PICTURE   **THE BIGGER PICTURE**

When you are planning to use moral or ethical debates or discussions in science, consider including cross-curricular links with other subjects. The obvious one would be by using drama and role-play techniques to develop children's speaking and listening skills, but you could also build in links to numeracy where evidence collected includes statistics and other numerical data, and ICT with research into the issues combining several strands by creating multimedia presentations to use in the role-play debate.

Teachers' questions

One of the most effective strategies for teaching primary science is for teachers to use a wide range of questions to prompt thoughtful responses in the children. Questions can be classified as open-ended or closed but not all questions fit neatly into these categories. Questioning is a skill which is learnt with practice and experience since it often involves 'thinking on one's feet'.

Teachers use open and closed questioning for a variety of reasons.

- To elicit children's ideas – *Can you tell me how you think the bulb lights up? How can you hear the sound made by the guitar string?*
- To help make connections between new and existing knowledge – *Now that you have seen the bread dry up on the window sill, how is this like the saucer of milk which we saw dry up the other day?*
- To highlight the steps in a causal sequence – *What has happened so that the condensation forms on the glass of cold water?*
- For discussion, prediction and explanation – *How do you think the flat piece of paper will fall? Why do you think the woodlice prefer to live in dark, damp places?*
- To focus children's attention on key science ideas – *Is the end of the tuning fork moving? Which kind of birds ate the breadcrumbs from the bird table?*
- To promote the links between science ideas – *If we have a lot of energy we can run faster. How could you give the car on the slope more energy and what would happen if you did?*

For more on questioning, see Chapter 7.

Consider the following strategies when using questions in science.

When talking to the whole class.

- Encourage all children to attempt an answer by asking a question, then waiting a while before choosing someone to answer it. Insist on hands up to achieve this.
- Encourage less motivated children by asking them directly what they think.

- Encourage more children to be involved in giving answers by giving no response after the first answer. This enables others to have a go before knowing which is the acceptable answer.
- Give praise for a correct answer or a partly correct one.
- Give praise for attempting to answer, even if the answer is wrong.
- Try to include all children at some time. Do not get into a habit of only asking children near the front or who are more keen to answer than others.
- Ensure you get a gender balance in the children you ask.

When talking to a group.

- There is more potential here for allowing children to provide answers without hands up but don't be afraid to impose this rule if necessary.
- Give children a chance to respond to each other's answers in a group situation.
- Encourage children to ask questions of one another if possible.
- Take a less formal approach in a small group, if possible, to make it easier for children to contribute.

When talking to individuals.

- Put the child at ease as soon as possible. Use humour and encouragement to do this if necessary.
- Use facial expression and body language to encourage responses to questions.
- Don't fire off too many questions at once. Intersperse these with conversation and/or instruction.

Generally:

- Listen to a child's response carefully. Base your next question on what is said and your interpretation of it.
- Make the question easier if a child is struggling.
- Lead a child towards the required answer by asking a series of simpler questions.

Strategies for differentiating work in primary science

See Chapter 7 for more on differentiation.

Any group or class of children will be made up of a range of abilities, levels of attainment and personalities. Teachers will therefore always need to consider the differentiation of the work they provide for children. This will involve work which will stretch the more able as well as special tasks or support for the less able. Many teachers prefer to consider children in terms of attainment rather than ability, e.g. high and low attainers rather than more or less able. More able children are not necessarily high attainers, and vice versa, and the terms should be used with care. It is important to have high expectations of all children regardless of ability or attainment at all times. There are a variety of ways in which teachers can differentiate work in science, including by:

- preparing work at different ability levels for ability groups within the class;
- having mixed ability groups in which different tasks are allocated according to ability or preferred learning style;
- providing graduated material in which different abilities begin and finish at different points;

- providing extension work for more able children within the class, to be done once they have achieved the main task;
- providing extra teaching support for those who are less able and/or those who are more able – for instance by paying more attention to such a group or allocating a classroom assistant to the group;
- arranging for the more able children to work with the less able ones;
- expecting a different outcome from different groups – this works especially well when the task is an open-ended one such as a problem-solving exercise or a science investigation;
- using computer-aided learning which has differentiation built into it;
- modifying and using different resources for different groups.

It would also be a mistake to assume that a child who is of a certain ability in, say, maths or English, will be of a similar ability in all aspects of science. For instance, quite often teachers comment on the fact that children who are of an average ability in most subjects are often particularly able when it comes to some aspects of science, especially practical work. Also, children have different preferred learning styles as well as abilities. Thus, a child who prefers to express their scientific ideas verbally may not do so well in writing and vice versa. Therefore, teachers may want to differentiate the way children plan, execute and respond to work in science to suit such characteristics.

Strategies for organising teaching groups

There are a variety of ways of organising groups of children to do science in the classroom. Some examples include the following.

For more on organising groups, see Chapter 8.

- One group does science while others do other subjects – this enables more attention to be given to the science group.
- All children do the same activity as individuals/pairs/groups – this enables the teacher to talk to the whole class about work in progress.
- Use of the activity carousel in which groups do different science activities on a similar theme and then rotate round until all tasks are completed – this enables the use of scarce equipment and the build-up of a particular concept through many short activities.
- Each group does a different but related activity and provides feedback to the whole class – this enables an accumulation of experience within the class within a short time and encourages explanation and discussion.

Getting the most support from additional adults

In order to ensure that any additional adults support the children's learning in science effectively:

For more on briefing additional adults, see Chapter 5.

- make sure that you brief any other adult about the purpose of the lesson;
- check that any additional adults understand the science involved in the lesson;
- ensure that additional adults know exactly what they are expected to do, including any assessment that needs to be carried out.

A SUMMARY OF **KEY POINTS**

> **All activities should lead to a progression in children's learning.**

> **Teachers should have an understanding of the ideas that children bring to science lessons and be prepared to encourage children to challenge and reconstruct these.**

> **Practical activities are an effective way of learning science but not all science can be taught in this way.**

> **Non-practical activities such as the use of secondary sources of information, analogy and role play can play an important part in teaching science.**

> **The precise use of language plays an important part in the effective learning of science.**

> **There are a variety of strategies for differentiating work in primary science.**

> **There are a variety of strategies for effective teacher exposition.**

> **There are a variety of strategies for effective classroom organisation.**

> **There are a variety of strategies for employing teacher's questioning.**

M-LEVEL EXTENSION > > > > M-LEVEL EXTENSION > > > >

Using the 'In the Classroom' examples and the suggested Further Reading below, consider again the strategies discussed in this chapter.

a) Are they suitable for particular aspects of science, or equally as useful for Life processes and living things, Materials and their properties and Physical processes?

b) How could you plan for progression in each strategy across the age and ability range of the phase in which you are teaching?

c) Ask your mentor if it is possible for you to arrange to observe experienced class teachers and the subject leader for science when they have planned lessons using one or more of the strategies covered in the chapter and to discuss these with them after the observations.

REFERENCES REFERENCES **REFERENCES** REFERENCES REFERENCES

Harlen, W. (2004) *The Teaching of Science in Primary Schools*. London: Fulton.

Peacock, G. A. (1999) *Teaching Science in Primary Schools*. London: Letts.

Skamp, K. (2004) *Teaching Science Constructively*. London: Harcourt Brace.

FURTHER READING FURTHER READING **FURTHER READING** FURTHER READING

Arthur, J. *et al.* (2006) *Learning to Teach in the Primary School*. Abingdon: Routledge.

Harlen, W. (2000) *Teaching, Learning and Assessing Science 5–12*, 3rd edn. London: Paul Chapman. Chapter 7 on the teacher's role provides good advice on a variety of strategies to be used in the classroom to promote effective learning in science.

Hayes, D. (2000) *The Handbook for Newly Qualified Teachers: Meeting the Standards in Primary and Middle Schools*. London: Fulton.

Peacock, G. A. (2002) *Teaching Science in Primary Schools*. London: Letts. Each chapter gives examples of suitable teaching strategies for primary science topics.

QCA/DfEE (1998, with amendments 2000) *Science: a Scheme of Work for Key Stages 1 and 2 – Teacher's Guide*. London: QCA. The teacher's guide provides some brief advice on classroom strategies for implementing the scheme.

7
Planning

Introduction

Planning for science lessons will involve a number of features which are perhaps not found in other subjects. An important feature for children of primary school age will be a varied approach which is often practical in nature and relates the science ideas to their everyday experiences. Good planning will take account of current ideas about how children learn science as well as how they might learn to 'think like a scientist'.

One of the greatest challenges to teachers of primary science is to understand for themselves the scientific ideas they wish to teach. They have to know well beyond what they would expect their children to learn. This can pose particular problems for those who teach towards the end of Key Stage 2, where the science becomes ever more sophisticated. There are no easy solutions to this dilemma. Very competent teachers towards the end of their career will still be learning more about even the simplest science ideas that they teach.

Primary Science: Knowledge and Understanding (2011) *in this series tackles the issue of science background knowledge for teachers and provides a sound basis for learning more.*

Teachers who plan practical activities for children should be wary of the seemingly simple activities described in many books and worksheets. What appears to be straightforward on the printed page can often become difficult in practice. It is therefore essential when planning activities to try them out practically beforehand. Some of the best primary science planning happens in the kitchen at home! Simply *writing* a lesson plan for a practical task does not guarantee success!

The following sections outline some of the major considerations to be made when planning primary science.

Levels of planning

This chapter is closely linked to the section on differentiation strategies in Chapter 6.

Planning for science is the work you do before the lesson so that it runs smoothly, addresses the learning needs of the children and fits into a wider programme of learning for those children. There are different levels of planning for different time scales:

- long-term planning;
- medium-term planning;
- short-term planning.

Long-term planning

At some time, a broad overview or long-term plan of the learning in science for each child throughout the primary school years will be required. In a single primary school, this is often prepared by the science subject leader or a small team of teachers working in conjunction with the whole staff. Long-term plans are often referred to as schemes of work. Schemes of work are based on a school policy for science, which is often produced by the science subject leader. It is debatable which of these two documents actually comes first. In an ideal world, policies are established and everything else builds on that firm basis. In reality, a clear policy will often emerge from current practice, which may be based on a partly formed scheme of work.

A scheme of work has a number of features which will help in the formation of effective plans for teaching.

- It will map out briefly what each year group within a school should be studying in science over each term.
- It may show broad learning objectives and learning outcomes for each block of work.
- It may suggest where there should be a particular emphasis in science investigations.
- It may refer to useful resources, including the publications available in school.
- It may suggest cross-curricular links.

There are a number of major functions which a scheme will enable the science subject leader and other teachers to perform. A scheme can be used to ensure:

- coverage of the current curriculum requirements for science;
- that there is a broad and balanced set of experiences in science;
- that there is suitable progression and continuity in science learning;
- that some science topics are revisited within the primary years;
- that unnecessary repetition is avoided.

Medium-term planning

Medium-term plans enable teachers to gain an overview of a single block of work which may last from two or three hours to a whole term. A medium-term plan contains more detail than a typical scheme of work and will most often be used by a year group team of teachers. A medium-term plan maps out briefly the stages that the children will move through, perhaps on a week-by-week basis. It contains more detailed learning intentions and learning outcomes which might be achieved over the period of time to which it applies.

Teachers use a medium-term plan as a guide and base their individual lesson plans on this. Features of a medium-term plan include:

- any preparation or prior work that needs to be done before a unit is started;
- more specific learning objectives and learning outcomes to be achieved over a period of time;
- links with programmes of study from the current science curriculum;
- brief descriptions of each learning experience, perhaps with a suggested time scale for each;
- cross-curricular links;
- assessment opportunities;
- examples of children's activities;
- an indication of the resources available for the topic.

An exemplar scheme of work for primary science (QCA/DfEE, 1998, with amendments 2000) provided an overview of suggested work in science as a series of units for the whole primary school and yet contained sufficient detail to be used as a single topic planning document. Some schools used this scheme alone; others used it to supplement another scheme. Some schools are now replacing or adapting science schemes to emphasise features such as skills development, creativity and cross-curricular links.

All long-term and medium-term plans should be the subject of constant review by the school. The plans are therefore working documents which are adaptable enough to be changed to keep up with current developments in the subject.

Short-term planning

Lesson plans provide guidance for the teacher and communicate to others in the school exactly what is intended to happen in a single lesson. Trainees on school experience will be expected to produce detailed lesson plans in order to communicate their ideas to a wide audience including their class teacher. It is often the first kind of detailed planning that a trainee or newly qualified teacher is involved in because the broader planning will have already been done by others.

At the heart of a good lesson plan are the learning intentions for that lesson. These are often, but not always, the starting point for detailed short-term plans. Some schools and local authorities refer to learning objectives as learning intentions to distinguish them from learning outcomes.

In the past, many teachers were tempted to begin with identifying an exciting science activity which was certain to interest a group of children, then work 'backwards' to identify the learning objectives which fitted this. Placing the learning intentions second, however, meant that often the same ones were covered more than once and some important objectives remained neglected. The formation of a scheme of work and topic plan first means that it is essential to begin planning lessons with learning objectives drawn from the long-term and medium-term plans.

Figure 7.1 provides examples of broad learning objectives drawn from the QCA/ DfEE Scheme of Work (1998, 2000) together with examples of more specific learning objectives suitable for related lesson plans.

Unit of work	Relevant programme of study statement from National Curriculum for science	Broad learning objective from medium-term plan	Example of a related learning intention for the lesson
Ourselves	Life processes and living things: 4a Pupils should be taught to recognise similarities and differences between themselves and others.	Children should learn that there are differences between humans.	By the end of the lesson, most children should be able to collect and record data on the eye colour of a group of children.
Plants and animals in the local environment	Scientific enquiry: 2a Pupils should be taught to ask questions and decide how they might find answers to them.	Children should learn to turn ideas of their own, about what plants need to begin to grow, into a form that can be tested.	By the end of the lesson, most children should be able to make a list of the things they think a plant needs to grow.
Circuits and conductors	Materials and their properties: 1c Pupils should be taught that some materials are better electrical conductors than others.	Children should learn which materials are better conductors than others.	By the end of the lesson, most children should be able to test these six things for conductivity: aluminium foil, steel key, paper, copper ornament, pencil lead, wooden rod.
More about dissolving	Scientific enquiry: 2c Pupils should be taught to think about what might happen or try things out when deciding what to do, what kind of evidence to collect, and what kind of equipment and materials to use.	Children should learn to make predictions about what happens when water from a solution evaporates and to test these predictions.	By the end of the lesson, most children should be able to predict what will happen to the salt solution and should be able to link this prediction to a previous experience.

Figure 7.1 The relationship between programmes of study, objectives for the topic and learning intentions for the lesson

Learning intentions should be written in a particular way. They should not be broad descriptions of what the teacher will do in the lesson but rather statements of what the teacher intends the children to learn by the end of the lesson. It is often a good idea to begin:

> *By the end of the lesson, most children should be able to...; know...; understand ...*

This will ensure that you will describe what children are intended to achieve in terms of what they should be able to do, to understand or to know.

There need only be about two or three learning intentions for each lesson, and these will provide the main focus of the work. Often, more than is required will be achieved. Learning objectives are best shared with the class on the board, inter-active whiteboard or a flipchart. Learning objectives can be differentiated to cater for higher- and lower-achieving children, as can be seen in the example lesson plan below. The importance of clear learning objectives will become apparent when planning to assess the children's understanding.

For more on assessment, see Chapter 9.

PRACTICAL TASK PRACTICAL TASK PRACTICAL TASK PRACTICAL TASK

Devise some suitable learning intentions, appropriate for individual lessons, which would be related to each of the following extracts from National Curriculum Science (1999).

Key Stage 1: Materials and their properties

1b   Children should be taught to sort objects into groups on the basis of simple material properties (for example, roughness, hardness, shininess, ability to float, etc.).

Key Stage 1: Scientific enquiry

2h   Children should be taught to make simple comparisons and identify simple patterns or associations.

Key Stage 2: Physical processes

3e   Children should be taught that sounds are made when objects (for example, strings on musical instruments) vibrate, but that vibrations are not always directly visible.

Increasingly, teachers are using learning intentions/objectives to involve the children in self- or peer-assessment. To assist with this, and with their own assessment of the learners, some teachers use success criteria. These are often two or three short statements of exactly what the teacher will be looking for in the learners' work or behaviour, for example:

Learning intention/objective – by the end of the lesson, most children should be able to collect and record data about the eye colour of a group of children.
Success criteria – collect data about the eye colour of children in Year 1; record data about eye colour on a tally chart.

A typical lesson plan format

It is often helpful to write lesson plans into a formatted page. Each lesson is different, however, and different amounts of detail will be required in each section. It can also be helpful to use a series of headings and write as much as is necessary under each. A typical lesson plan for science might use the headings shown in the example below.

TITLE _____

Date/Class and year group/Number of children/Duration

Curriculum and scheme of work references
Refer to the appropriate programme(s) of study and units of work. These may include references to the level descriptions where appropriate.

Relationship to the school's long-term and medium-term plans
Which parts of the school's plans does this lesson address? Does it form part of a sequence?

Children's previous experiences
What have most children experienced on a similar subject?

Learning intentions
By the end of the lesson:
– most children should be able to...; know...; understand...
– in addition, more able children should be able to...; know...; understand...
– less able children should be able to...; know...; understand...
(Include reference to skills, knowledge and understanding, children's attitudes, cross-curricular skills such as literacy, numeracy and ICT, children's spiritual, moral, social and cultural development, etc.)

Lesson sequence
Introduction – set the scene – show how you will relate the work to children's everyday lives.
Child activity – how will the lesson develop?
Conclusion – summarise the key points of the lesson.

These three basic stages of a lesson might be described in terms of:
(a) What the teacher will do.
(b) What the children will do.
(c) How the children and the classroom will be organised.

Classroom organisation and management
Reference to use of space, whole class/group/individual teaching, approximate timing of lesson phases/activities, role of teacher and other adults, allocation of resources, etc.

Differentiation strategies
How will you support those who are less able? How will you provide extension work for those who need it? Do individual children have special requirements?

Homework
Is this lesson a basis for setting homework? What would this be?

Assessment opportunities
Which learning intention(s) will form the basis for your assessment? Who will you assess? How will you assess? What evidence will you gather to support your assessment? How will you use your assessment in your future plans?

Recording and reporting
How will you record the results of your assessment? Which aspects of this lesson might be reported to others in the school, parents or the children themselves?

Health and safety
Are there any health and safety issues associated with this lesson?

Not all headings need to be used for every lesson plan. You may find it helpful to treat the above as a checklist so that no important piece of information is left out.

THE BIGGER PICTURE   THE BIGGER PICTURE   THE BIGGER PICTURE

When you are planning homework activities, they must, of course, be purposeful and either extend children's knowledge or understanding or consolidate a skill already learnt in school. You can involve parents in supporting their children's learning by sharing in advance the topics or aspects of science that you plan to study. This is helpful especially if you are asking children to undertake research tasks as they might plan a trip to their local library or want to supervise children's safe use of the internet.

Children's science ideas

Are you aware of the ideas currently held by your children concerning the science you are about to teach?

A great deal of research has been carried out into the ideas that children hold in some of the major areas of science, for example the Science Process and Concept Exploration or SPACE Project, the Children's Learning in Science Project (CLISP) and so on. The research demonstrated that children learn science by building on what they already know. This is the constructivist view of learning science. When planning a series of science lessons, it is often useful to elicit the children's ideas about the science they are to learn first and to consider these ideas carefully. This may have implications for the way a topic or an individual lesson is structured. For instance, before launching into a series of activities on the subject of forces it might be necessary to plan an elicitation task which enables children to express their own ideas. Further planning might then depend on the results of this exercise.

For more on this research, and how to build on it, see Chapter 4.

Lesson structure and pace

Have you thought about how you will adjust the structure and pace of the lesson to suit most of the children?

Lesson structure is made up of the variety of learning activities and the timing of each. It is important to plan for child experiences and teacher intervention. Different groups of children may have differently structured experiences in the same lesson. For instance, a simple structure for a class of Year 4 children might consist of:

- the lesson introduction 15 minutes;
- the main child activity 40 minutes;
- the lesson conclusion 10 minutes.

Many lessons, however, will not (and should not necessarily) conform to this pattern.

IN THE CLASSROOM

Miss Kaur wrote her learning intentions for a lesson on making electrical switches as follows.

By the end of the lesson, most children should:
- be able to find and rectify faults in a simple electrical circuit;
- be able to use two batteries in a circuit to light a bulb;
- know how a variety of handmade switches can be used to control electricity in a circuit.

The Year 3 class had already learnt in the previous lesson how to join a battery to a bulb to make it light and to a buzzer to make it buzz. Miss Kaur now wanted to teach the children how to overcome simple problems if their circuits did not work. She gave them each a troubleshooting table and demonstrated each fault to the children.

PROBLEM Circuit does not work because:	REMEDY
The bulb is loose in its holder	Screw it down
The bulb is broken	Replace the bulb
There is a loose or dirty connection to the bulb holder	Wiggle connector/clean connector
The crocodile connector is faulty	Replace connector with one which is known to work
The battery is loose in its holder	Wiggle the battery
One battery is the wrong way round	Turn the battery
The battery is dead	Replace with a battery known to be working
The buzzer is the wrong way round	Change the buzzer connections

The children were then allowed 10 minutes to make a simple circuit with two batteries in a holder and a bulb or buzzer. While they were doing this, Miss Kaur went around to each group and 'sabotaged' their circuits by unscrewing a bulb, changing one battery round or making some buzzer connections a little loose! She made it clear that it was the responsibility of the children to try to rectify any faulty circuits whenever they had problems.

The children were then asked to watch while Miss Kaur showed them how to make some simple switches for their circuits using aluminium foil, paper clips and card. A worksheet was produced with a number of different switch designs on it and the children set to work to make as many as they could in the time available.

At the end of the lesson some children were able to demonstrate their completed switches in their bulb or buzzer circuit.

The structure for Miss Kaur's lesson enabled her to make two 'introductions' – one to discuss troubleshooting and the second to demonstrate making switches. She used a brisk pace for the first part because she knew the children would come back to the idea of troubleshooting throughout their work on electricity. She allowed more time for making a variety of switches, thus enabling the work to be differentiated. This was achieved by increasing the difficulty of each switch described on the worksheet and expecting all children to make the easier ones first.

Introducing a science lesson

Do you begin your science lessons in a variety of ways?
Consider beginning the lesson by adopting one or more of the following:

- recapping on the previous lesson;
- reading a story to set the scene;
- showing a picture or poster;
- discussing an artefact you have brought to school, e.g. a push-along toy for a topic on forces;
- watching a short video or using a CD-ROM presentation;
- referring to an out-of-school visit the children have recently made;
- taking the children for a short trip around the school, for instance to look at the use of building materials;
- demonstrating a scientific phenomenon such as a paper helicopter dropped in front of the children;
- asking children for their experiences of a relevant topic such as examples of materials which are waterproof;
- eliciting children's ideas about the way shadows are formed during the day by asking them to make a poster;
- asking children to begin by exploring, for instance, a collection of magnets to find out whatever they can.

Children's activities

Do you vary the kind of learning activity experienced by the children?
In the development of the lesson, children might spend most of their time on activities such as:

- careful observation of an object or collection of things such as seeds or rocks;
- drawing, discussing or describing in writing the things they have observed;
- doing something practical by following instructions which might be presented verbally, on a workcard, worksheet or the board/IWB;
- using role play to help understand a science idea;
- planning and carrying out a science investigation;
- recording and interpreting scientific results;
- reporting their findings and conclusions to others – verbally, in writing, through drawings, etc.;
- completing a worksheet by responding with written answers and drawings, etc.;
- completing a quiz;
- using secondary sources of information to learn new ideas, e.g. using posters, reference books, CD-ROM, TV or video clips or the internet;
- asking questions.

EMBEDDING ICT EMBEDDING ICT **EMBEDDING ICT** EMBEDDING ICT

When you plan to use secondary sources, remember that there are colleagues in school who can advise you on what is available. Check the school's long- medium-term planning in case it identifies available resources, talk to the science subject leader, the class teacher or your mentor and check the VLE wherever possible, before deciding whether you will need to source any additional materials yourself. Always pre-view video clips and TV programmes, simulations and models and use programs hands-on to ensure that they are appropriate for your requirements and suitable for the children's needs.

Practical activities

Do your children develop their science ideas through simple practical activity?

There is a strong tradition that learning in science in the primary school is often achieved through practical work. Not all ideas are best learnt through practical science activities, but the nature of the subject is such that these often play a more important role than in many other subjects. Practical work requires special planning skills which consider the distribution and maintenance of resources, time management and lesson structure. There are no hard and fast rules, but the following are worth thinking about before planning a practical lesson.

- Do the children understand how to behave in a practical activity situation? Do they need special instructions in this respect?
- How much time in the lesson should be spent on practical work?
- Is it appropriate to use teacher demonstration for part of the time?
- Should you control the practical work by having individual children demonstrate to the class?
- Should children work alone or in groups?
- How should learning be consolidated during or after the practical activity?
- Which kind of practical work best suits the learning objectives?
- How should instructions be provided so that the children can work independently?
- Should extension work be provided for those who finish early?
- What support will some children need?
- How will equipment be cleared away?
- What should be done about faulty equipment during the lesson?
- How should children record their ideas and procedures?
- How should the learning be summarised?
- What assessment opportunities will arise?

Preparing to ask questions

For more on asking questions, see Chapter 6.

Do you promote effective learning by asking a variety of types of question of individuals, groups and the whole class?

One of the most important and effective ways to teach primary science is to ask appropriate questions of the children. It is often possible to plan a set of key questions which might be used as a stimulus to asking further questions during the lesson. Asking a string of appropriate and challenging questions 'off the top of one's head' is not easy and this is a skill which takes time to acquire. It helps to think of the different categories that teachers' questions might fall into and to plan to ask a variety of different types. Consider these examples on the subject of light and colour.

- Where does light come from?
- Can you name two things that give out their own light?

- Can you tell me some things you can think of which are shiny?
- Which is your favourite colour?
- Do you think you can see in the dark?
- What are the colours of the rainbow?
- How long does it take light to reach us from the Sun?
- What happens to light when it is reflected from a mirror?
- Do you think light can pollute the environment?
- How do you know the colours you see are the same as anyone else?

Some of these questions require a direct, factual answer. We could describe these as *closed questions*. Some questions, however, require a more expansive answer which may include personal opinion. These could be described as *open-ended questions*. Notice that some questions have neither a correct nor incorrect answer. You should plan to ask a variety of types of questions while avoiding the trap of asking only closed questions because these appear more 'scientific'.

Closed questions can:

- test specific recall;
- be used to assess knowledge, understanding and prior learning;
- require only a limited response (e.g. a one-word answer);
- sometimes be perceived as threatening to a child;
- possibly be avoided by a child who claims to be unable to answer.

Open-ended questions can:

- elicit a range of responses – some surprising!
- be less threatening than closed questions;
- elicit a longer, fuller and often more revealing response than closed questions;
- be less likely to be avoided by a child.

REFLECTIVE TASK

Consider the following questions a teacher might ask on the subject of electricity. Decide which questions are closed and which are open. Do all questions fit neatly into these categories? Which questions would you find threatening?

1. Can you think of some things which use electricity?
2. What do we call the thing we use in the classroom which gives us electricity?
3. What is inside a torch bulb?
4. If I tie a knot in a wire in an electrical circuit, what do you think will happen to the brightness of the bulb?
5. What name do we give to the pushing power of a battery?
6. Electricity can be used to make heat. What else can it do?
7. What do you think electricity is?
8. If electricity is like the flow of water in a pipe, how could we show the effect of putting a light bulb in the circuit?
9. Do you think living near electricity pylons is good for your health?
10. What would the world be like without electricity?

Differentiation

Is it necessary to differentiate by task or by outcome from some of the work in the lesson to cater for less able and more able children?

For more on differentiation, see Chapter 6.

Part of a successful plan for a science lesson will involve some suggestion for catering for different abilities or levels of attainment within a group or class. An example of such provision would be planned extension work for those who finish early. A different kind of extension may be devised for those who are particularly able. These children may not necessarily finish early but may need directing towards something more challenging. Less able children may need different instructions or different work to do. Alternatively, you may plan to provide extra help for them as they complete the work that most children will do. In some instances, work may be differentiated by accepting different outcomes from a task. This might be true for a group of children who are given an open-ended task of planning a simple investigation. The more able children might be expected to produce a more sophisticated plan. Some children may have preferred learning styles within a class, such as a preference for discussing ideas rather than writing them down.

THE BIGGER PICTURE   THE BIGGER PICTURE   THE BIGGER PICTURE

When you are planning differentiated activities, consider including differentiation for learning styles. Check if the school as a whole or the class teacher you are working with routinely uses a model of learning styles. There are several different theories; the most straightforward one that is often used in schools divides children into three different types of preferred learning style:

- visual;
- auditory;
- kinaesthetic (or tactile).

For more on the different teaching strategies that best suit each style, read Neil Fleming's (2001) book *Teaching and Learning Styles*.

Progression

Does the work you have planned enable children to move on from their current knowledge and understanding? Does it link to what they know already?

Children should progress in:

- their knowledge and understanding of science concepts;
- their ability to carry out scientific enquiry.

The issue of progression is often dealt with at the long- or medium-term planning stage. Broad features of progression in primary science include children moving from:

- using everyday language to increasingly precise use of technical and scientific vocabulary, notation and symbols;

I pulled the toy to make it move along ——➤ *A force was necessary to make it turn a corner.*

- personal scientific knowledge in a few areas to understanding in a wider range of areas and of links between areas;
  I know puddles dry faster in the playground when it is windy ——➤ *Puddles dry through evaporation in the same way that clothes dry on a line, cakes go stale and paint dries.*
- describing events and phenomena to explaining events and phenomena;
  The screwed-up piece of paper fell faster ——➤ *The screwed up piece of paper fell faster because it had less air resistance.*
- explaining phenomena in terms of their own ideas to explaining phenomena in terms of accepted ideas or models;
  I can see the clock by aiming my eyes at it ——➤ *Light from the bulb falls on the clock face and is reflected into my eyes.*
- participating in practical science activities to building increasingly abstract models of real situations.
  I investigated which kind of leaf the snail preferred to eat ——➤ *I understand the food chain, which includes the bird that eats the snail that eats the leaf.*

(Adapted from QCA, 1998, p. 6)

Progression within individual lessons can be as important as overall progression. Children should move towards the learning objectives of the lesson, beginning with what they already know and understand, and progressing towards new ideas. Teachers can facilitate this kind of progression by structuring their lesson into a series of 'digestible chunks' of learning. Every opportunity should be taken to challenge and extend the children's knowledge, understanding and skills (and not just simply keep them occupied) through questioning and directed tasks. The conclusion to a lesson is often a time when a teacher might push understanding to its furthest possible limits.

Assessment and recording

Have you thought about how you will know whether the children have achieved the learning intentions/objectives?
Planning for assessment is as important as planning what the children will do. All levels of planning should include details of where and how children will be assessed. See Chapter 9 for more details of this.

Further considerations in planning science lessons

Are there other aspects that should be included in lesson plans?
The following aspects should also be considered in lesson planning.

See Chapter 10 for how ICT might be used.

- Are there any links with other subjects that would enhance the science?
- How will these links be made?
- Is there potential for using ICT to support the science learning in this lesson?
- Is there a list of scientific and technical vocabulary which you would like the children to learn to use in this lesson?
- Have you made plans for the role of any other adults who you may have in the classroom, and how will you brief them?

Concluding the lesson

Do you vary the way in which you bring a lesson to a close?

Children are often willing to pay special attention towards the end of the lesson so this is a good opportunity to consolidate teaching. Also, this time gives you a chance to praise good work and reinforce good practices. You might conclude a lesson by:

- summarising the science the children have been learning;
- showing examples of successful work by some children or groups to celebrate or consolidate learning;
- extending the children's learning by asking suitable questions;
- collecting together the results of a class investigation;
- asking individuals or groups to explain some of their ideas or describe what they have been doing to assess levels of understanding;
- challenging some of the misunderstandings that may have arisen during the lesson;
- discussing what the children will be doing next in the sequence of lessons.

Evaluating the lesson

Do you evaluate the planned lesson effectively?

Part of a good planning strategy is to reflect on what happened in a lesson and record a written evaluation. This is not the same as the assessment of children's achievements. Child assessment, however, might be used to inform the lesson evaluation. It is a good idea to focus on different areas of the teaching in each evaluation in order to target specific aspects for improvement. Each evaluation might contain some or all of the elements shown in the exemplar pro forma below:

TITLE_____

Date/Class and year group/Number of children/Duration

Planning for the children's performance
Was the content of the lesson appropriate and stimulating for most children? Did the planned work stretch the children in their educational development?
How did the children respond to the lesson? (Refer to the learning intentions if this is appropriate.) What were the reasons for their response?
Did the children learn what was intended? Did any children fail to benefit (e.g. able children, average children, less able children, shy children, boys/girls, vulnerable groups, disruptive children, etc.)? If so why?

My performance as a teacher
What went well? Why was this so? What could have been improved? Did I work well with other adults in the classroom?

Points for future planning
What have I learnt about the subject content, the way to deliver this subject and the way children learn?

Targets from previous lesson evaluations
Have I achieved any targets that I have previously set myself?

Targets for the future
Set two or three personal targets to aim for when teaching in the future.

A SUMMARY OF **KEY POINTS**

> **Planning effective learning takes place at three different levels:**
>   – **long-term plans that are often made by the subject leader;**
>   – **medium-term plans that are often made by the subject leader and year group team;**
>   – **short-term 'lesson' plans that are made by the class teacher.**
> **Long-term and medium-term plans may be combined into one scheme of work.**
> **Lesson plans should begin with clear learning intentions/objectives.**
> **Key elements in a lesson plan include:**
>   – **finding out children's existing science ideas;**
>   – **adopting a suitable structure and pace;**
>   – **designing suitable activities for the children;**
>   – **asking effective questions;**
>   – **effectively differentiated work and well structured progression for all children;**
>   – **assessment opportunities.**
> **Teachers should reflect on the success of their lessons by evaluating them and setting targets for future lessons.**

M-LEVEL EXTENSION > > > > M-LEVEL EXTENSION > > > >

Building on your studies in previous chapters as well as this one, talk to children about their ideas concerning different scientific concepts and undertake research into different models of learning styles. How could these influence your short-term planning for individual lessons or sequences of lessons?

REFERENCES REFERENCES **REFERENCES** REFERENCES REFERENCES

QCA/DfEE (1998, with amendments 2000) *Science: a Scheme of Work for Key Stages 1 and 2*. London: QCA.
QCA (1998) *Science: a Scheme of Work for Key Stages 1 and 2 – Teachers' Guide*. London: QCA.

FURTHER READING FURTHER READING **FURTHER READING** FURTHER READING

Arthur, J. (2006) *Learning to Teach in the Primary School*. Abingdon: Routledge.
Harlen, W. (2000) *Teaching, Learning and Assessing Science 5–12*, 3rd edn. London: Paul Chapman Publishing. A good general introduction to planning science lessons. Particularly strong on assessment.
Harlen, W. (2004) *The Teaching of Science in Primary Schools*. London: Fulton.
Hayes, D. (2000) *The Handbook for Newly Qualified Teachers: Meeting the Standards in Primary and Middle Schools*. London: Fulton.
Peacock, G. A. (1999) *Teaching Science in Primary Schools*. London: Letts. Provides a wealth of ideas to use in the classroom in a structured science programme.

8
Classroom organisation and management

Introduction

The organisation and management of primary science can vary as much from class to class as from school to school. Such variation simply reflects how different teachers (and trainees) respond to the different circumstances they find themselves in and have to deal with on a day-to-day basis (e.g. the age and previous science experiences of the children, the differing teaching environments), how each school chooses to present and teach the science curriculum as a whole (e.g. fully integrated topics, science-based topics or as discrete subjects) and the diverse nature of schools themselves (e.g. when built, their location, their physical layout).

RESEARCH SUMMARY RESEARCH SUMMARY RESEARCH SUMMARY **RESEARCH SUMMARY**

Most authors agree that while there is no one way of organising and managing all lessons within primary classrooms, there are several features of classroom organisation and management which are associated with good practice and which are recognised to contribute towards effective teaching and learning as a whole (e.g. Hayes, 2003, 2004; Kelly, 2006). Within the context of primary science, a range of strategies have been described in more detail. In brief, successful science lessons are most likely to result from a mixture of whole-class, group and individual teaching, having only a small number of groups engaged in science activity at any one time, selecting an appropriate classroom layout and making good use of space, good time management, good behaviour management, and considering carefully the role of the teacher and other adults. The use of ICT and display are also important.

In this chapter, organisation and management are considered in terms of:

- teaching the whole class, groups and individuals;
- classroom layout and space;
- time management;
- behaviour management;
- availability and choice of appropriate resources, including ICT;
- the use of display;
- the role of the teacher and other adults.

Teaching the whole class, groups and individuals

Organising and managing the whole class, groups and individuals depends to a large extent on how as well as what you want children to learn. Throughout the primary years, increasing use is being made of science lessons which are planned and taught within well defined lesson parts. These parts commonly include an introduction, a main science activity (or activities) and a plenary/review. The use of lesson parts is both efficient and effective and ensures that a mixture of whole-class, group and individual teaching takes place. Lane *et al.* (2005) give a good account of ways in which children can be grouped in science.

See Chapter 6 for more on teaching strategies.

Whole-class science teaching can eliminate some of the repetition involved when working only with groups. Whole-class science teaching is particularly suited to:

- the introductory and plenary parts of lessons (e.g. setting the scene, instructing, demonstrating, explaining, reviewing, sharing, discussing, concluding work);
- questioning and eliciting ideas (e.g. finding out what children already know and the nature of their errors and misconceptions);
- sharing science texts and reading together (e.g. 'big books', researching in libraries);
- emphasising matters associated with health and safety (e.g. breakages, spillages, accident and emergency routines);

- educational broadcasts (e.g. using television to access or extend science that cannot be undertaken in the classroom).

When teaching the whole class, firm rules need to be established for determining who is allowed to talk and when (e.g. hands in the air, listening to and respecting the views of others). Moyles (2007) suggests that rules need to be established quickly, with fairness and consistency. Teachers and trainees also need to consider their position and movement within the classroom or teaching area (e.g. fixed or mobile), teaching style (e.g. formal, informal) and how to keep the children involved and motivated (e.g. interaction). Transitions from whole-class teaching to group work and back need to be carefully controlled to avoid lost time and disruption.

Group work is perhaps best suited to the main phases of science lessons and would normally, though not necessarily, be expected to follow on naturally from whole-class introductions. Grouping allows children to work together collaboratively in organised social settings. The number of groups within a class of children and the number of children within each group may vary but should always be easily manageable. Four or five groups, each with between five and seven children, are particularly common. Group composition is usually determined on the basis of attainment, friendship, control or a mixture of all of these. Typical arrangements include:

- all groups working on the same science activity at the same time – a variation of whole-class teaching (easy to monitor, intervene and control, requires high resource availability, allows for differentiation by task or by outcome);
- all groups working on different science activities and sharing findings – (reasonably easy to monitor, intervene and control, requires moderate resource availability, promotes communication skills);
- all groups working on different science activities in rotation – sometimes referred to as a carousel of activities (good control of rotation essential, requires moderate resource availability, activities should not be linked in terms of sequence or progression, time on activities restricted);
- only one group working on science – rest of the class working on other things (useful if science activity needs close supervision or attention, requires low resource availability; monitoring behaviour and quality of work in other groups may be challenging).

It is essential when working with groups not to be over ambitious at the expense of depth of coverage of the science being taught or to allow the science to lose its distinctiveness, particularly if teaching within fully integrated topics. Careful planning for all forms of group work is required.

All teachers and trainees should spend quality time, no matter how short, with each and every child throughout the school day. In science, individual teaching is most likely to occur during the main activity phases of lessons. Working with individuals is particularly useful for monitoring progress, sorting out difficulties perhaps associated with using equipment, collecting, interpreting, analysing and representing data, and clarifying ideas. Working with individuals also helps to strengthen relationships and to build bonds. Children not only receive individual help, but they can also ask questions and converse in less formal ways without feeling intimidated or humiliated by peers.

Reflect critically and analytically on your own experiences of teaching science to a whole class, groups and individuals. In terms of efficiency and effectiveness, consider the advantages and disadvantages of each way of working. Would efficiency and effectiveness have been affected if you had worked any differently? Can you think of any aspects of science teaching which are better suited to one particular way of working (e.g. effective exposition including questioning, instructing, demonstrating and explaining, skills, concepts and attitudes development, using models and analogies, particular aspects of scientific enquiry)?

Classroom layout and space

For most teachers and trainees, science will usually be taught in a classroom or other teaching area. Sometimes, science is taught elsewhere (e.g. in the hall, in corridors, in a special science room, a kitchen or outdoors). Factors which may affect or influence the layout of classrooms and other teaching areas include the size and shape of the space available, the age and number of children present, the amount of furniture required, and the location of exits, storage and resource areas, sinks, boards and windows. Teachers usually begin by experimenting with different layouts to find out which works best for most teaching situations – not only science. Freedom of movement around the room and clear lines of sight are essential. Moving furniture for every lesson is not always possible or convenient. This takes time and can cause disruption. Four common layouts include:

- simple groups;
- horseshoes;
- rows and columns;
- mixed arrangements.

Many classrooms and teaching areas include workstations. Science workstations help contextualise work, focus children's attention and concentration, and provide ready access to resources.

REFLECTIVE TASK

Examine each of the classroom layouts in Figures 8.1 to 8.4. Consider the similarities and differences between these and what you have seen in schools. In terms of advantages and disadvantages, which layouts are best suited to whole-class, group and individual teaching or all three? Which layouts make best use of space and allow for good eye contact and overall class control? Which layouts are best suited to teaching science in Key Stage 1 and Key Stage 2 or both? Does layout matter if teaching about life processes and living things, materials and their properties, and physical processes? Which layout do you prefer and why? Given the opportunity, and the year group of your choice, how would you arrange your own classroom?

THE BIGGER PICTURE   THE BIGGER PICTURE   THE BIGGER PICTURE

When you are planning to reorganise your classroom, keep in mind the requirements of health and safety (of both children and adults, including your own!). Take advice from other teachers

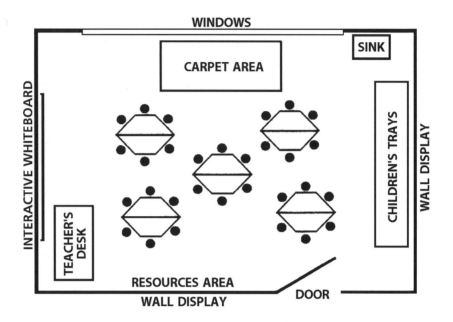

Figure 8.1 Simple groups

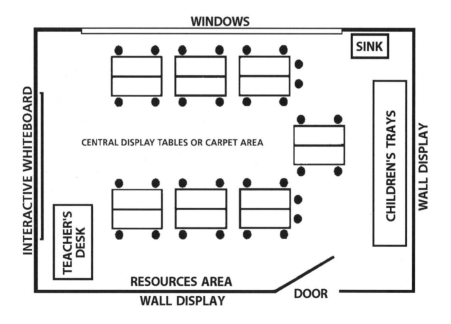

Figure 8.2 Horseshoe

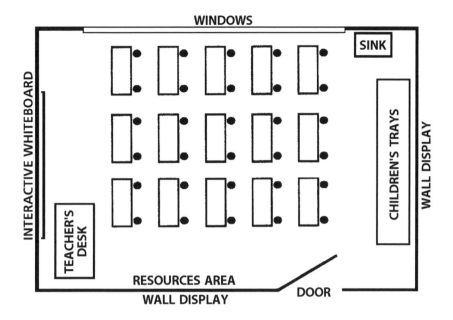

Figure 8.3 Rows and columns

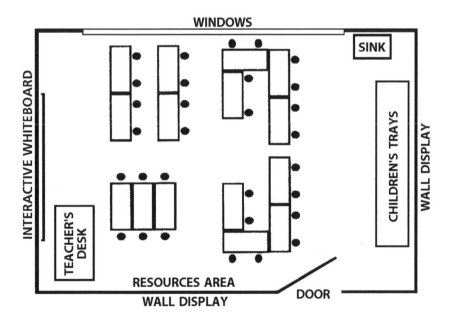

Figure 8.4 Mixed arrangement

and senior staff, and the site manager or caretaker. Also, bear in mind the safeguarding agenda, for example don't cover external or corridor windows and glass panels completely with displays or posters. It is considered safer for children to have sightlines available, and it is also in your best interests as it reduces the possibility of any misunderstanding or false allegations. Always adhere to school policies in these areas.

Time management

Time is a valuable commodity in primary science, a commodity which cannot easily be replaced. This is worth remembering when some science investigations or experiments go badly wrong and need to be redone, or take much longer to complete than originally anticipated. The time allocated to individual science lessons in primary schools can vary from as little as 30 minutes to as much as 90 minutes or more per week. It is important to realise, however, that within that time, children usually have to be settled, lessons have to be introduced and brought to a satisfactory conclusion, activities have to be completed, classrooms or teaching areas have to be tidied up, and equipment has to be returned. Careful planning, keeping an eye on the clock and experience really do help. It is also worth remembering that not all of the time spent in school is spent teaching. Avoid wasting time.

- Always be thoroughly prepared and ensure that the resources needed for the lesson are readily available.
- Ensure that the teaching methods selected match the requirements of the lesson.
- Maintain pace, focus and direction and do not become distracted.
- Be aware of the rate at which different children work and what they can and cannot do.
- Work to ensure that children spend as much time as possible on-task.
- Note any times when other teaching and non-teaching duties may interfere with teaching time.

Don't forget that time will also be required for marking, assessing and discussing work, revision and keeping records, some of which will be done with the children, for example self- and peer-assessment and giving feedback.

Behaviour management

Maintaining order and a good standard of discipline during science lessons is particularly important, not only to secure effective teaching and learning, but to ensure that all matters associated with health and safety are observed and that preventable accidents and injuries are avoided (Adams, 2009). Establishing clear rules and routines and setting high expectations for behaviour is essential. Children can misbehave and become disruptive during science for any number of different reasons, for example where:

- lessons are badly organised and managed (e.g. unsatisfactory classroom layout and use of space, noise levels too high, transitions disorganised, inadequate time to complete activities, unsuitable resources);
- groupings are inappropriate;
- the science is too easy or overly challenging;
- intervention and supervision are inadequate.

Signs of misbehaviour and disruption can soon become all too obvious. Children may talk out of turn, not pay attention, stray around the room, interfere with others' work, or fail to get on with and finish their own work. Ways of dealing with minor incidents include:

- use of eye contact;
- signalling gestures;
- varying the use of your voice;
- issuing specific commands;
- praise and positive reinforcement of appropriate behaviours.

Serious breaches of discipline may require more serious action to be taken. This should be undertaken in accordance with the school's behaviour policy. Whatever happens, always stay calm and composed and try to avoid confrontation at all costs. Do not be afraid to stop a lesson, particularly if you feel that health and safety are compromised.

Availability and choice of appropriate resources, including ICT

It is often said that high quality science teaching requires high quality resources. While this is undoubtedly true, high quality resources can only improve the quality of science teaching if they are readily available, appropriate for the task at hand and used effectively. While many commercially available science resources are more sophisticated now than ever before, they can also be very expensive. Many high quality science resources are simply collected by teachers or made when needed. Science resources include:

- items of scientific equipment (e.g. for measuring length, time, volume, mass, temperature and force, magnifiers, magnets, mirrors and lenses, batteries and bulbs, weather boards, models, safety goggles, disposable gloves and so on);
- a range of different materials (e.g. paper, textiles, wood, glass, plastics, etc.);
- science schemes including worksheets and workcards (e.g. Nuffield, Ginn, Bath);
- educational broadcasts (e.g. BBC, Channel 4, Teachers' TV);
- audio tapes, video clips, slides (e.g. commercially available);
- books (e.g. fiction, non-fiction, 'big books');
- ICT (e.g. word processors, CD-ROMs, databases, sensors, control technology devices);
- museums and science centres (e.g. education officers, lending facilities, visits);
- the local environment (e.g. school grounds, parks, beaches);
- guest speakers (e.g. parents, professional scientists).

Most of the commonly used science resources listed above are to be found in schools on a permanent basis. Typical arrangements include:

- central storage;
- topic boxes;
- class sets.

In large primary schools, Key Stage 1 and Key Stage 2 resources may be held separately.

Central storage systems keep all science resources together in a designated area. This might be a storage cupboard, part of a corridor or an unused classroom. Storage space is kept to a minimum, access is possible to all teachers at all times, and resource duplication is minimised, thus saving money for a wider range of resources. Central storage systems located far from classrooms and other teaching areas may, however, inhibit spontaneous work in science, particularly if teachers are unable to retrieve what they need and school rules require that children may only do so under supervision. Careful curriculum planning is also essential to avoid demand for the same resources at the same time. Topic boxes store everything needed for a particular topic, and may be kept together in central storage or, alternatively, in classrooms where they are most needed. Resources may be retained for the duration of a topic and any additional items collected along the way can be added to the collection. Topic boxes may require some resource duplication, perhaps at considerable cost. Topic boxes may also have to be shared between two or more parallel year groups. Class sets ensure that each class has what it needs to deliver the science being taught throughout the year. This allows for immediate and unrestricted access. However, resource duplication may be excessive and maintenance costs may be high.

EMBEDDING ICT EMBEDDING ICT EMBEDDING ICT EMBEDDING ICT

For more information on using ICT in science, see Chapter 10.

The use of computers and other forms of ICT is integral in primary schools. It is important that teachers are familiar with the hardware and software available to support the teaching and learning of science. Typical computer arrangements include:

- a separate computer room or suite;
- a cluster of computers (usually along corridors or in specific locations/hubs around the school);
- a set of laptops available for use in the classroom;
- one or two PCs in a classroom or other teaching area.

A computer room or suite is useful for whole-class teaching but access may be restricted to particular times. Children may lose sight of the science in their enthusiasm to use the computers. A cluster of computers is useful for group teaching but access may still be restricted. The location of the cluster may require the teacher or a teaching assistant to be out of the classroom or away from the teaching area for long periods of time. Single computers are useful for up to three children at one time. Access is almost unrestricted. Monitoring children's work and ensuring equal access is important.

PRACTICAL TASK PRACTICAL TASK PRACTICAL TASK PRACTICAL TASK

At the start of your next school placement, familiarise yourself with the range of science resources available and how these are organised and managed. In terms of the science to be taught, consider carefully what you need, what is available and, as a result, what you must obtain. What are the advantages and disadvantages of the resource organisation and management strategies adopted by the school?

Teaching children how to identify the resources they need, how to look after them and use them safely, and how to collect and return them by themselves is also important. This not only develops a sense of responsibility; it helps lead children towards becoming independent learners. It also saves teaching time. Just as it is important to plan for progression in scientific knowledge and understanding, it is also important to plan for progression in the selection and use of resources. By the end of Key Stage 1, for example, children should be familiar with and able to use an appropriate range of scientific equipment with some help from their teacher. By the end of Key Stage 2, children should be familiar with and able to use a wider range of scientific equipment by themselves. They should know that different pieces of scientific equipment have specific functions, recognise the importance of precision and accuracy in measurements and, where a selection of items is available, be able to choose the best for the task at hand.

EMBEDDING ICT EMBEDDING ICT EMBEDDING ICT EMBEDDING ICT

The use of computers requires teachers to be aware of security (e.g. personal databases), unrestricted internet access (e.g. unsuitable websites), and health and safety (e.g. keeping loose cables stored tidily, prohibiting food and drink being brought near computers, avoiding adverse lighting and screen glare, ensuring proper tables and chairs are used, avoiding repetitive strain injury, and so on).

Don't forget that one of the best, and certainly the biggest, resources is outdoors. *Primary Science Review* 91 devoted an issue to this subject, with articles by Dillon (2006) and de Boo and Shaw (2006). Catling (2006) gives useful advice on safety and organisation outdoors.

IN THE CLASSROOM

Rajinder was coming to the end of her first term of teaching with a class of seven-year-olds. When she first arrived, she naturally fell into the system operated by the school and the other year group teachers around her. Despite successful teaching practices in similar environments during training, things were not going well, particularly with science. The quality of work being produced by the children was lower than expected and Rajinder was worried. She raised this issue with her mentor who very wisely suggested that she consider other ways of working but which didn't disrupt the practice of her colleagues. The first thing Rajinder did was to reduce the number of groups she taught from six to four, to be made up on the basis of broad levels of attainment rather than friendship. Next, she identified an area of the open-plan teaching area which she could easily transform into a science workstation. The school was well resourced for science and kitting out the workstation did not present any particular problems. Finally, Rajinder shifted emphasis from only teaching science in groups to operating a lesson phase system similar to that used in literacy and numeracy. After only a few weeks, she noticed considerable improvements. Not only was the quality of science much better (teaching and learning), but the children also seemed to be enjoying their new way of working, and Rajinder felt happier too! Her mentor, an experienced teacher, gave her the opportunity of sharing her changes during an after-school meeting. Her other year group colleagues were so impressed they agreed to try them out for themselves.

The use of display

Displays are used in all primary schools. Whatever the age of the children involved, all displays should be big, bold, colourful, informative and interactive. They should also be kept simple and uncluttered. Good science displays:

- recognise and reward children's efforts;
- develop self-esteem;
- help generate more purposeful and stimulating classrooms and teaching areas in which to work;
- can be used to revise, consolidate and extend learning;
- promote the development of children's scentific knowledge and understanding and allow them to build on and refine their ideas.

Presentation should always be of the highest quality. Titles and other headings should be clear and legible (e.g. handwritten, cut from paper or card or word-processed) and all work should be single or double mounted where appropriate. Wall-mounted displays are probably the most common and easiest to produce. Displays on table tops and other work surfaces, together with mobiles suspended from ceilings, are also popular. Displays should be changed on a regular basis in order to reflect the science work being undertaken and should always be taken down or replaced when looking tired. Creating science displays can be time-consuming and resource-intensive. Forward planning and a sketch are recommended. Children should be fully involved in the overall design and preparation of displays where possible. Mini-displays can also be used to remind children of scientific vocabulary, how to write scientific reports, how to use scientific equipment properly, the steps in a scientific investigation or experiment, and health and safety.

Figure 8.5 Interactive displays enhance teaching and learning

The role of the teacher and other adults

At the start of every science lesson, it is essential that teachers know exactly what their role and the roles of any other adults present will be. While working with other adults in the classroom or teaching area is now commonplace, it is worth remembering that all teachers are ultimately responsible for everything that goes on at all times. Other adults may include:

For more on briefing other adults, see Chapter 5.

- parent helpers present on a purely voluntary basis;
- fully trained and paid learning support and teaching assistants;
- other teachers, including special educational needs co-ordinators or those supporting children with English as an additional language;
- guests and visitors.

Decisions need to be made in advance about when different individuals become involved in the lesson, the extent to which they participate, which groups or individuals they work with and for how long, and what other work they are doing if not directly involved in teaching. This will help to avoid ambiguity, minimise duplication of effort and prevent confusion among children. Establishing positive working relationships and clear lines of communication is essential. Do not expect any help with science teaching beyond the level of expertise or training of the other adults present.

A SUMMARY OF **KEY POINTS**

> **There is no one way of organising and managing all science lessons within primary classrooms or teaching areas.**

> **The organisation and management of primary science can vary as much from class to class as from school to school.**

> **Organisation and management is as much about how as about what you expect children to learn.**

> **Primary schools make increasing use of well-defined lesson phases incorporating a mixture of whole-class, group and individual science teaching.**

> **Other features of classroom organisation and management which are associated with good practice and which are recognised to contribute towards effective science teaching and learning include appropriate classroom layout and good use of space, good use of time, appropriate selection and use of resources, good behaviour management, effective use of ICT and displays, and the carefully considered role of the teacher and other adults.**

M-LEVEL EXTENSION > > > > M-LEVEL EXTENSION > > > >

As you make the transition from your course of initial teacher training into your induction period and NQT year, you should continue to monitor, evaluate, reflect on and review how you organise and manage your classroom to ensure effective teaching and learning. Do not assume that your initial preferred mode of working will turn out to be the best or last the pace. Analyse your practice carefully and always be prepared to adjust it accordingly. Take some time with your induction tutor or mentor to visit other classrooms and observe how things are set out and done there. Walk around the school and note how classroom and more public displays are constructed and presented. Much of what you do may turn out to be beyond your control (due to school policy or because they are curriculum driven), but always be prepared to try out new ways of working where you can. Use the reference and further reading publications to research these aspects in more detail to inform your practice.

REFERENCES REFERENCES **REFERENCES** REFERENCES REFERENCES

Adams, K. (2009) *Behaviour for Learning in the Primary School*. Exeter: Learning Matters.

Catling, S. (2006) Planning for learning outside the classroom, in Arthur, J. (ed.) *Learning to Teach in the Primary School*. Oxford: Routledge.

De Boo, M. and Shaw, B. (2006) Science at the Seaside. *Primary Science Review*, 91, 18–20.

Dillon, J. (2006) Education! Education! Out! Out! Out! *Primary Science Review*, 91, 4–7.

Hayes, D. (2003) *Planning, Teaching and Class Management in Primary Schools*. London: Fulton.

Hayes, D. (2004) *Foundations of Primary Teaching*. London: Fulton.

Kelly, P. (2006) Organising your classroom for learning, in Arthur, J. (ed.) *Learning to Teach in the Primary School*. Oxford: Routledge.

Lane, R., Coker, J. and McNamara, S. (2005) Getting to grips with group work. *Primary Science Review*, 90, 25–27.

Moyles, J. (2007) *Beginning Teaching, Beginning Learning*. Maidenhead: Open University Press.

FURTHER READING FURTHER READING **FURTHER READING** FURTHER READING

Ward, H., Roden, J., Hewlett, C. and Foreman, J. (2005) *Teaching Science in the Primary Classroom*. London: Paul Chapman.

9
Assessment, recording and reporting

Professional Standards for the award of QTS

Those awarded QTS must have a secure knowledge and understanding of science that enables them to teach effectively across the age and ability range for which they are trained. To be able to do this in the context of assessment, recording and reporting in science, trainees should:

Q1 Have high expectations of children and young people including a commitment to ensuring that they can achieve their full educational potential and to establishing fair, respectful, trusting, supportive and constructive relationships with them.

Q3(a) Be aware of the professional duties of teachers and the statutory framework within which they work.

(b) Be aware of the policies and practices of the workplace and share in collective responsibility for their implementation.

Q4 Communicate effectively with children, young people, colleagues, parents and carers.

Q11 Know the assessment requirements and arrangements for the subjects/ curriculum areas in the age ranges they are trained to teach, including those relating to public examinations and qualifications.

Q12 Know a range of approaches to assessment, including the importance of formative assessment.

Q13 Know how to use local and national statistical information to evaluate the effectiveness of their teaching, to monitor the progress of those they teach and to raise levels of attainment.

Q22 Plan for progression across the age and ability range for which they are trained, designing effective learning sequences within lessons and across series of lessons and demonstrating secure subject/curriculum knowledge.

Q26(a) Make effective use of a range of assessment, monitoring and recording strategies.

(b) Assess the learning needs of those they teach in order to set challenging learning objectives.

Q27 Provide timely, accurate and constructive feedback on learners' attainment, progress and areas for development.

Q28 Support and guide learners to reflect on their learning, identify the progress they have made and identify their emerging learning needs.

Q29 Evaluate the impact of their teaching on the progress of all learners, and modify their planning and classroom practice where necessary.

The guidance accompanying these standards clarifies these requirements and you will find it helpful to read through the appropriate section of this guidance for further support.

Introduction

Assessment has four main functions. These are to inform:

- the current teacher about planning for future teaching;
- children about their own learning and progress;
- subsequent teachers and the school about children's learning, progress and attainment;
- parents about their children's learning and progress.

The first two of these functions are generally achieved through formative assessment, while the second two are generally the result of summative assessment. Formative assessment gives evidence that helps to develop a programme of teaching, while summative assessment gives evidence of the child's level of attainment at a particular time.

Formative assessment

Before discussing formative assessment in the classroom, we need to briefly consider diagnostic assessment as a type of formative assessment. Diagnostic assessment is most often done by experts in a particular field and focuses on the analysis of particular learning difficulties. Summary results of diagnostic assessment by specialist agencies are often used by teachers to help inform their work. Diagnostic assessment is not covered in this chapter as it is rarely possible for a teacher running a busy classroom to have the time or experience to do this effectively.

Formative assessment, which is carried out by class teachers, informs the next steps in teaching children and also informs children about their own progress. Teachers need to know what their children understand and how they learn before they begin to teach them. This avoids wasteful repetition or teaching children ideas and skills that are beyond their ability to learn. Once this information has been established, teachers need to find ways to measure how much of what they teach results in children learning. This has two benefits:

- the rate of progress through the work can be differentiated to suit the class or parts of the class;
- the teacher can evaluate the effectiveness of different teaching strategies.

For more on differentiation, see Chapters 6 and 7.

IN THE CLASSROOM

Here, we look at how one teacher approached the teaching about the uses of materials at Key Stage 1. It is an example of using formative assessment.

When analysing the work, Majid intended to include the teacher-identified key vocabulary, concepts and skills that the children needed. He recognised that in this area of work it is important that children know the difference between an object and the material from which it is made. It is not the intention of this book to suggest that teachers should involve themselves in exhaustive testing before any teaching takes place. This would be counterproductive and waste valuable time. However, Majid carried out a quick revision assessment in order to ensure that the children gained the maximum benefit from the work in hand. He used a simple table to gain valuable evidence on which to base decisions about the best teaching approach:

Object	Material it is made from

To reinforce the point and assess present levels of understanding, Majid asked the question in a slightly different way. The table below asks children to say why three objects were made from a selection of materials.

Object	Material it is made from	Why was that material chosen?

After clarifying any issues arising from this introductory work and identifying those children who needed some help with basic concepts, Majid moved on to the main objective: why do we use a variety of materials for specific uses? The children had a number of actual objects to look at, handle and discuss and then used similar ways to record their ideas. Examination of the children's answers gave Majid clear information about how well the children understood the lesson about the use of materials. In addition, the level of the answers gave a clear indication about the success of his teaching approach.

Formative assessment strategies

There are many useful strategies for carrying out formative assessment in primary science. In addition to being useful for assessment purposes, many of these techniques can be used to vary the style of recording that children are asked to carry out. Formative assessment techniques include:

For more on learning intentions/ objectives, see Chapter 7.

1. modelling;
2. drawing explanatory diagrams and pictures;
3. labelling diagrams;
4. talking about concept cartoons and pictures;
5. simple structured tests;
6. word association;
7. more complex concept maps;
8. hot-pen writing;
9. practical skills tests;
10. observing investigative work;
11. children's peer- and self-assessments.

Modelling

Modelling done by children can give their teachers an impression of their level of understanding. For instance, young children given a lump of modelling clay can use it to show their idea of the shape of the Earth.

Drawing explanatory diagrams and pictures

Drawing explanatory diagrams and pictures can be a highly effective way to assess children's understanding. For example, children can draw a picture of what they think happens to the Sun at night. If the task is set with only a little guidance about the style and content of the drawing, the way in which the child represents what happens to the Sun at night gives useful information about how to teach the child.

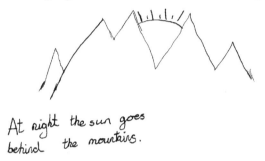

At night the sun goes behind the mountains.

Figure 9.1 Nine-year-old's drawing of the Sun behind mountains

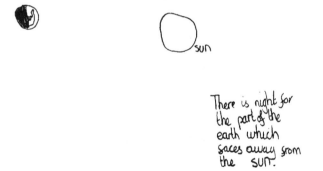

sun

There is night for the part of the earth which faces away from the sun.

Figure 9.2 Drawing of the Earth and Sun in orbit, by a nine-year-old child

For instance, a child who shows through their drawing that they think the Sun goes behind the clouds, hills or mountains at the end of the day (see Figure 9.1) needs quite different teaching to the child who appreciates that the spinning of the Earth causes the apparent motion of the Sun (see Figure 9.2). It is important to present tasks like this as fun – with a genuine and stated interest in finding out what the children already think.

In other contexts, Key Stage 1 children can be asked to draw a number of sound sources, while Key Stage 2 children can be asked to draw how they think sound reaches their ears. This style of assessment is very effective for delving into children's conceptual understanding and is likely to yield useful information about how to proceed (see Figure 9.3).

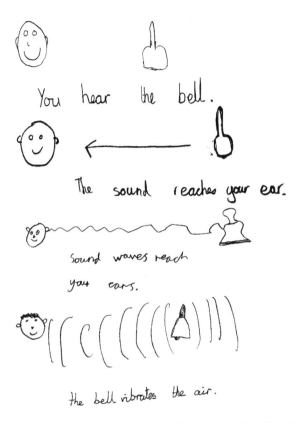

You hear the bell.

The sound reaches your ear.

Sound waves reach your ears.

the bell vibrates the air.

Figure 9.3 Four possible ways in which sound reaches our ears. The first drawing is by a six-year-old child, the middle drawings by eight-year-olds and the last drawing is by a nine-year-old.

Labelling diagrams

Completing partly drawn diagrams and drawings can be a useful and efficient way of assessing what children know. Here, the child has to add labels to a drawing, for example a diagram of a plant. This mode of assessment is highly efficient in discovering a child's level of factual knowledge. Another context in which this might be used could involve the child being asked to show in a partially completed drawing how the light from a lamp helps a person to see an object which also appears in the drawing.

For more on children's scientific ideas, see Chapters 4, 5, 6 and 7.

Talking about concept cartoons and pictures

Talking about concept cartoons and pictures is a fruitful way for children to begin to debate their ideas. This style of assessment encourages children to see that science can be a matter of opinion and that evidence is needed to support ideas. In a relaxed classroom environment, children are likely to voice their ideas about what they think is going to happen, and the listening teacher, while they may not gather a huge amount of information about an individual child, will gain a clearer understanding of the ideas of the class as a whole. There is a rich source of such material in Naylor and Keogh (2000). For instance, one of the many concept cartoons shows several children gathered around a snowman debating how best to keep him frozen (see Figure 9.4).

What do YOU think?

Figure 9.4 An example of a concept cartoon

One child asserts that the snowman will be best in a coat. Children reading the concept cartoon can be encouraged to join in the debate. There is no set right answer (as whether or not a coat will help depends on the air temperature and amount of sun) but the ideas offered by the children will give their teacher useful information about the correct level of work about melting and freezing.

Simple structured tests

Simple structured tests using pencil and paper or completed on-screen are a useful source of assessment information. At the start of a unit of work in science many schools give their children a simple structured test. The tests help clarify what the children already know and what they need to know. The essence of a test like this is that it must be easy to administer and mark. If the child does the marking and they do the same or a similar test at the end of the period of study, they are likely to appreciate how much they have learnt.

Below is the outline of a test that might be useful to a Year 2 or Year 3 teacher about to start on some work about electricity. In this test the diagrams have been replaced by brief descriptions in order to save space.

Question	Drawing	Level of the question and comments
What will happen when the switch is turned on?	Drawing of simple circuit with a switch.	Level 1: Change results from action.
What will happen when the torch is switched on?	Drawing of torch.	Level 1: Change results from action.
Which bulb will be brightest?	Simple torch bulb circuit and table lamp.	Level 2: Compare devices in electrical circuits.
Which of these bulbs will glow brightest?	Drawing of two circuits with bulbs, one with two batteries and similar one with only one battery.	Level 2: Compare devices in electrical circuits.
Why do you think that motor A will turn faster than motor B?	Drawing of two circuits with motors, one with two batteries and similar one with only one battery.	Level 3: Cause and effect.
Which of these bulbs will not glow? Explain your idea.	Drawing of two circuits with bulbs, one with a break in the circuit and the other complete.	Level 3: Link cause and effect in simple explanations.
Finish off this sentence to show what you know about electricity: If there are a lot of batteries in a circuit...	No drawing.	Level 3: Beginning to make generalisations about physical phenomena.

To devise these tests, the teacher has looked at the National Curriculum level descriptions for Physical processes (DfEE/QCA, 1999) to get the right level.

Word association

Word association is the simple precursor to the use of concept maps. These allow children to record what they know rather than be reminded about what they don't know. Word association and concept maps ask children to draw up a plan showing how they link different areas of knowledge. It makes concrete the connections we make between apparently disparate aspects of understanding. There are several levels to using concept maps, with word association being the most basic. For example, write the word 'animal' – what do you associate with that word?

animal – dog

animal – cat

animal – boy

More complex concept maps

More complex concept maps involve asking children how they connect a wider range of words. Young children or lower-attaining older children can be given pictures and large arrows to arrange on the floor or a large table. Give a few large words already written – for example, Earth, Sun, Moon, orbit, star, planet, and a selection of arrows. The children can arrange these on the floor and talk about the connections they are making. They can indicate connections by writing connecting words or sentences along the arrows.

Giving the children several words already written on a piece of paper along with small pieces of paper with words that they might connect to the words on the larger sheet is a development from the pictures and large arrows. For instance, the words 'plant, 'pine tree' and 'cactus' are written on a sheet of paper and the following words are written on small pieces of paper:

hot, dry habitat
cold, snowy habitat
trunk
leaves
stem
spiky
thick

The children are invited to stick the words onto the page and then draw arrows connecting the words. The arrows can be labelled to indicate the way in which they are connected (see Figure 9.5).

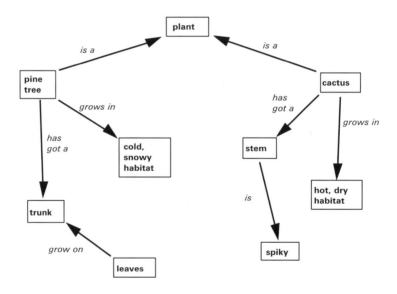

Figure 9.5 An example of a concept map

For more able or older children, two main words, for example Moon and Earth, can be written on the board. As a class, other words can be discussed which might be associated with the two main words. List these down the side: planet, star, orbit, seasons, day, night, spin, turn, eclipse, month, day, year, tilt and axis. Invite the children to construct their own concept diagrams before and after a unit on the Earth and beyond. Allow more able children greater freedom by omitting the list of words and invite them to link words of their own.

Kitchin (2000) researched the use teachers make of concept maps. He mainly looked at secondary students but his detailed analysis is highly relevant for all primary school teachers. Kitchin discusses the importance of allowing children to own their maps and for teachers not to become obsessed with the correctness of the map. Kitchin suggests that focusing on the correctness of a map turns the process into 'transmission teaching' but acknowledges that the idiosyncratic maps produced by children are difficult for teachers to interpret.

Kitchin suggests that maps tend to fall into three pattern categories:

- Spoke – which Kitchin suggests represents the National Curriculum structure;
- Chain – which Kitchin suggests represents a non-integrated lesson sequence;
- Net – which Kitchin suggests represents meaningful learning.

The author goes on to analyse the finer details of concept maps, looking in particular at the quality of the links between concepts. The strain of scoring and assessing the maps in quantitative ways is discussed in the paper. Two examples of concept maps flowing from the word 'flower' are used to illustrate the idea that children will produce quite different and idiosyncratic maps if they have different overall ideas. In the cases quoted, one of the children constructed a map based on plant reproduction while another child produced a map concerned mainly with the interaction of flowers with insects.

Hot-pen writing

Hot-pen writing is usually much enjoyed by children. Invite them to write and draw all they know about the topic you are about to begin. Give them a very tight time schedule, such as ten minutes. Again, after studying the topic, ask the children to repeat the exercise and compare the two versions. This provides invaluable information for evaluative assessment.

Practical skills tests

Activities that test skills can give very clear indications of children's abilities in the area of scientific enquiry. For example, before the class carries out an investigation into the rate of cooling of water in different containers, it may be useful to assess whether they can read the types of thermometers you have in the classroom. If the vast majority can already do so, then moments have been wasted but, if many children are still unsure, revision of reading scales would be productive. One extremely simple way to test this skill is to show a thermometer to the children as they leave the classroom and ask each one to whisper the reading into the teacher's ear as they go out to break.

Observing investigative work

Assessing classroom work is a routine part of the teacher's role. How teachers respond to children's work makes important differences to the way that children perceive their achievement. Many children feel that teachers judge neatness, spelling and punctuation rather than content. It is fairly easy to home in on these features as they are relatively easy to assess compared with aspects of content which may require longer comments. Perhaps we should aim for quality of comment rather than frequency.

The comments written by teachers on children's work can have important effects. Gipps *et al.* (2006) give ten principles, first articulated by the Assessment Reform

Group, which should be adhered to when giving oral or written feedback. However, while positive comments are welcomed by children, they find that simple comments like 'Good' are of little use to them. Gipps *et al*. suggest that, above all, children want to receive constructive guidance about how to improve. Children want to know what they can do to improve the content of their work and not simply their presentation. Comments should set targets for children. Teachers who set simple attainable targets for children in the context of comments on work encourage children to believe they can gain greater success in their work. Given achievable yet challenging targets they can become children who are motivated by a desire to learn rather than a desire to be seen to do well. Comments that contain targets are likely to lead children to believe they can improve. These might include:

> 'I liked the way you wrote about your results – can you explain a little more about what you did?'
> 'Your graph was easy to read – can you make the title describe more accurately what the graph is about?'
> 'The labelled drawing of your flower had four labels on it. Can you add two more?'

Overall and Sangster (2006) make it clear that feedback is often conveyed with subtle non-verbal clues. Feedback is a powerful tool that teachers use for comment on work or behaviour or for formative guidance on the content of work.

When observing investigative work, the judgements made about children's work should, as often as possible, be ipsative, i.e. the work should always be compared with the standard that the individual produced previously rather than being compared to the class as a whole. The latter may lead to the less able becoming disheartened and the able children becoming complacent.

REFLECTIVE TASK

Read Chapter 10 in Oliver (2006) and use the suggestions for assessment at the end of each chapter. The suggestions offered in the chapter are not focused exclusively on the National Curriculum and can be used in a range of situations. The chapter focuses strongly on creativity and open-ended tasks. The example of assessing 'detective work' gives an unusual slant on this.

- Are the children interested and involved throughout the activity?
- Do the children show confidence when faced with unexpected results?
- Do the children co-operate with others during the investigation?

Devise some similar assessment questions of your own, linked to your learning intentions/ objectives, to use informally to help you to judge the progress of individuals or the success of a session.

Children's peer- and self-assessment

Children's own assessment has a key role in the development of the child as a learner. Gipps *et al*. (2006) see that one of the main reasons for children to be fully involved in the assessment of their own learning is that they will shoulder more of the responsibility for their own learning if they are involved in assessing their own progress. They discuss the central role of open questioning in encouraging children to offer their own opinions about their own work and the work of others. It is important to avoid asking too many closed questions where only one response is deemed to be correct.

Harten and Qualter (2004) suggest that it is extremely important for children to know the purposes of their work. Furthermore, they believe that self-assessment reduces the need for teacher assessment and helps children to take ownership of their learning.

In order to assess their own progress, children need information about what they are learning. In particular, they need to know the objective for a particular session and also the context in which the learning takes place. So, for instance, it is useful to know the objective for a particular lesson, for example 'We are going to learn the names of the parts of the flower', or even more helpfully perhaps: 'Flowers are the way in which some plants make seeds, so we are going to see which parts are involved in doing that and give them names.'

One strategy that may help children's self-assessment is to ask them what they would like to know about the topic they are about to study and how they might do this. For example, they could draw up a simple table:

I want to know	I might be able to find out by
How many ants there are in a nest	Looking it up in a book
What snails like to eat	Giving them different foods to choose from
How many legs a caterpillar has	Looking at one through a lens or finding pictures of them

Try the following strategies for involving children in peer- and self-assessment.

- Give the children small attainable targets to aim for in their learning and ask them how far they feel they have met them.
- Tell the children about the learning intentions for the session.
- Emphasise the learning that children have achieved rather than the things they do not know by using assessment strategies such as concept maps (see above).
- After the learning, ask children to list what they have learnt and to compare their concept maps from the beginning of the work with those from the end.
- Help children to see where their learning fits into the bigger picture.
- Involve the children in constructively assessing each other's work.

THE BIGGER PICTURE   THE BIGGER PICTURE   THE BIGGER PICTURE

When you are planning opportunities for peer- and self-assessment, remember that these strategies are also useful in other areas of the curriculum, including maths and English. You may use talk partners in literacy and other subjects, and many teachers build on these relationships to extend them to peer assessment. However, it is important that any feedback shared between children is constructive and you will need to have agreed guidelines for the children to work to.

Recording formative assessment

Recording formative assessments must involve as little work as possible for the teacher. A simple grid such as the one below will suffice for most everyday purposes. Avoid rewriting the children's names each time by compiling a simple mark book.

	Objective: to be able to distinguish magnetic from non-magnetic materials 20/9/11	Objective: to draw up a table showing magnetic and non-magnetic materials 23/9/11	Objective: to be able to design a test to test the relative strength of three magnets 27/9/11	Objective:
James Brown	1 a	2 a	2 b	
Karen Gee	1 c	1 b	1 c	
Ahmed Khan	1 a	1 a	3 b	
etc.				

Key
Level of ipsative achievement (i.e. how did they do relative to their usual standard?)
1 = high   2 = average   3 = poor

Level of criterion-referenced achievement (i.e. did they achieve objective?)
a = high   b = average   c = poor

In addition to this everyday record, teachers may keep slim files of some assessed work to compare standards with other teachers. They would also find it helpful to refer to a summative record of each child in the class. A record such as the one below helps in several ways.

- It is easy to compile using information from the everyday record.
- It indicates which children are making rapid or slow progress relative to the class as a whole.
- It indicates the level of the majority of the class.

	Level 1				Level 2				Level 3				Level 4				Level 5			
	AT 1	AT 2	AT 3	AT 4	AT 1	AT 2	AT 3	AT 4	AT 1	AT 2	AT 3	AT 4	AT 1	AT 2	AT 3	AT 4	AT 1	AT 2	AT 3	AT 4
James Brown	x	x	x	x	x	/	/	/	/											
Karen Gee	x	x	x	x	x	x	x	x	/	/	x	/								
Ahmed Khan	x	x	x	x	x	x	x	x	x	x	x	x	x	x	x	/				
etc.																				

Key: A line to show some understanding and a cross to show mastery of content. In the actual document a summary of the content is shown. In this extract, Ahmed is assessed as having achieved all of Level 3 and is near to achieving Level 4.

ICT can be used not only as a direct support to children's learning through planned activities, but it can also help teachers with their ongoing assessment for learning, their regular record-keeping and in reporting on children's progress to their parents. There are commercially available programs to support all of these applications, but many local authority admin systems now have data-handling functions to track individual pupils' progress, their attainment as measured through summative assessments, and other important data such as attendance records.

Reporting to parents

Reports to parents should contain as much information in as straightforward a style as possible. Some schools only employ longhand reports while others use a word-processed pro forma that can convey a great deal of information very efficiently. Read this report which is based on the child 'David' whose work can be seen in SCAA (1995).

> *David is an observant boy who is keen to discuss his ideas with others. He enjoys science and puts forward his own ideas and suggestions about why things happen. He is methodical in his written work and conveys his ideas clearly. He can use diagrammatic symbols for electrical components and he labels diagrams with care. He records his findings in tables where appropriate. He links cause and effect in many of his explanations, which can sometimes be full and give a clear picture of what he has done. On occasions, David is able to make generalisations from his work but he needs help to make connections between related phenomena and he needs assistance to interpret his results correctly. There are occasions when David gives partial explanations in his work and he sometimes misses out important details. David makes interesting links between things he has noticed in the environment and his work in science. Overall, David's work is satisfactory for this class but slightly below average for his year group nationally.*

PRACTICAL TASK PRACTICAL TASK PRACTICAL TASK PRACTICAL TASK

Consider the following.

- Is the last sentence true?
- Is it useful?
- Will the child be disheartened?
- Does it imply the class is below national standards?
- Would the parents be pleased if the writer had been less blunt?

Do you think it is possible to design a pro forma which can convey all this information more efficiently? How would you rewrite the report to be more helpful for David's parents?

Summative assessment

National tests are compulsory in science at Year 6 but are otherwise optional at Years 2–5. The optional tests may be used to demonstrate how much children have progressed relative to the previous year. The national tests are criterion-referenced in that all children who answer to a set standard for a particular level can be awarded that level irrespective of the number of others who achieve the same standard.

From September 2008, children in the EYFS have been assessed against a comprehensive Early Years Foundation Stage Profile, the main purpose of which is to provide Year 1 teachers with reliable and accurate information about each child's level of development as they enter Key Stage 1. As mentioned at the start of this book, it is possible that assessment arrangements could be changed following reviews of both the Early Years and the primary National Curriculum.

Summative judgements are made by teachers at several points in the school year, most notably in compiling end-of-year reports. However, the highest status and highest stakes summative judgements are undoubtedly the national end-of-key-stage tests. These tests can be useful in evaluating the effectiveness of different schools and standards nationally but they are not helpful as formative assessments as they:

- give a single number summarising a range of data;
- provide very little in the way of detail for future work;
- use only one style of collecting data;
- must have a clear paper evidence trail against which to judge the findings.

RESEARCH SUMMARY RESEARCH SUMMARY RESEARCH SUMMARY **RESEARCH SUMMARY**

All writers emphasise the importance of discussion in the assessment process. However, it is clear that it is not always easy to elicit the sort of talk that will help teachers to interact with their children's ideas. Simon *et al.* (2008) describe a research project in which large puppets were used to help children articulate their ideas and teachers to alter the style of their discourse. They report that the level of reasoning increased as the puppets facilitated problem-posing and productive cognitive conflict for the children.

A SUMMARY OF **KEY POINTS**

> **Formative assessment helps teachers plan the next steps for their children's learning.**

> **Assessment also provides teachers with information about their own effectiveness.**

> **Assessment can frequently take place in the context of everyday classroom activities.**

> **There is a range of available strategies for formative assessment.**

> **Children value feedback from their teachers that helps them to make progress.**

> **Record-keeping need not be overly time-consuming.**

> **Reports to parents need to be honest and encouraging.**

> **Summative assessment is subject to much debate and may change in format following review.**

M-LEVEL EXTENSION > > > > M-LEVEL EXTENSION > > > >

Find out about Assessing Pupils' Progress (APP) (The National Strategies, 2010) and in particular about how it relates to science. Look in detail at the five assessment focuses for science and reflect on how APP could influence your assessment for learning (AfL). Will there be a knock-on effect on how you plan curriculum activities for children? Should there be?

REFERENCES REFERENCES **REFERENCES** REFERENCES REFERENCES

DCSF (2008) *Statutory Framework for the Early Years Foundation Stage*. London: DCSF.

DfEE/QCA (1999) *Science: the National Curriculum for England*. London: HMSO.

Gipps, C., Pickering, A., McCallum, B. and Hargreaves, E. (2006) From 'TA' to Assessment for Learning: The impact of assessment policy on teachers' assessment practice, in Webb, R. (ed.). *Changing Teaching and Learning in the Primary School*. Maidenhead: Open University Press.

Harten, W. and Qualter, A. (2004) *The Teaching of Science in Primary Schools*. London: David Fulton.

Kitchin, I. (2000) Using concept maps to reveal understanding: a two-tier analysis. *School Science Review*, 81(296), 41–46.

The National Strategies (2010) *Assessing Pupils' Progress: A teacher's handbook*. London: DCSF.

Naylor, S. and Keogh, B. (2000) *Concept Cartoons in Science Education*. Cheshire: Millgate House.

Oliver, A. (2006) *Creative Teaching: Science in the Early Years and Primary Classroom*. London: David Fulton.

Overall, L. and Sangster, M. (2006) *Assessment: A Practical Guide for Primary Teachers*. London: Continuum.

SCAA (1995) *Exemplification of Standards: Science*. London: SCAA.

Simon, S., Naylor, S., Keogh, B., Maloney, J. and Downing, B. (2008) Puppets promoting engagement and talk in science. *International Journal of Science Education*, 30(9), 1229–48.

FURTHER READING FURTHER READING **FURTHER READING** FURTHER READING

Arthur, J. (2006) *Learning to Teach in the Primary School*. Abingdon: Routledge.

Black, P., Harrison, C., Lee, C., Marshall, B. and Wiliam, D. (2003) *Assessment for Learning: Putting it into Practice*. Maidenhead: Open University Press.

Naylor, S. and Keogh, B. (2004) *Active Assessment*. London: Fulton.

10
Using ICT in science

Introduction

Information and communication technology (ICT) is a core subject in the National Curriculum. It is not only taught as a subject in its own right, but it is also used to support and develop other subjects. The Qualifications and Curriculum Authority (QCA) produced an optional scheme of work (QCA, 2003) that showed how ICT can be used to help children learn in a variety of contexts. You may find that schools that you work in use this scheme or elements of it in their planning for ICT.

Uses of ICT in science

There are four main aspects of ICT which are taught at Key Stages 1 and 2. These are:

- finding things out;
- developing ideas and making things happen;
- exchanging and sharing information;
- reviewing, modifying and evaluating work as it progresses. (QCA, 2003, p.6)

Not all ICT involves the use of laptop and desktop computers. It can include the use of digital cameras, scanners, mobile phones and a vast range of other hand-held devices.

Becta was the government agency leading a national drive to ensure the effective and innovative use of technology through learning in our schools. Becta closed at the end of March 2011.

RESEARCH SUMMARY RESEARCH SUMMARY RESEARCH SUMMARY **RESEARCH SUMMARY**

ICT is a fast-moving area and a good source of recent research can be obtained from the archive of the former Becta website (now held at www.education.gov.uk).

Becta believed that ICT can make science more interesting, authentic and relevant, allowing more time for observation, discussion and analysis. Using ICT increases opportunities for communication and collaboration.

The main uses of ICT in science are as follows.

- Writing reports using word processing, incorporating spreadsheets, photos and drawings.
- Storing and manipulating information using spreadsheets and databases.
- Using sensors to detect and measure changes in the environment (datalogging).
- Obtaining information from electronic sources such as CD-ROMs and websites.
- Using images from digital cameras, film and scanners.
- Creating illustrations using draw or paint programs.
- Communicating through e-mail and networks.
- Presenting information in charts, tables and via presentation software such as PowerPoint.
- Simulation software.
- Digital microscopes.
- Interactive whiteboards to provide a platform for a wide range of educational software.

Writing reports

Children may need to be helped to concentrate on the science they wish to report rather than the superficial appearance of details such as the size and colour of a font. Teachers can use simple pro formas to help children plan their scientific investigations. For Key Stage 1 children, an example might be:

We want to find out ...

We will do it like this ...

We think this will happen ...

What actually happened was ...

Our explanation is ...

These entries may be saved by the teacher as a template. More detailed pro formas can provide differentiated support for children of different abilities.

THE BIGGER PICTURE **THE BIGGER PICTURE**   THE BIGGER PICTURE

When you are planning cross-curricular work, consider how writing frames developed for literacy sessions could help to support less able or less confident writers in your class, including those children with additional and special educational needs. For example, a frame to structure and record research could be useful when undertaking a project on healthy lifestyles that includes work in science (healthy eating – food as fuel), personal, social and health education (good mental health, the dangers of alcohol, drug and substance abuse) and PE (the benefits of health-related exercise). If considering such a project, you may find it helpful to visit the MEND programme website (see www.mendprogramme.org). MEND (Mind, Exercise, Nutrition ... Do it!) is a joint initiative between local authorities, NHS primary care trusts and other partners to work together to combat the causes of child obesity, and successful elements include the healthy schools programme, extended schools activities and school sport partnerships.

Storing and manipulating information

A spreadsheet is a matrix of cells that you can use as a table of results. Once it has been filled in, you can sort results, carry out calculations and quickly draw a variety of graphs and charts. This allows time for the interpretation of the graph rather than spending time on the mechanics of drawing it. Microsoft's Excel is a sophisticated computer spreadsheet. At first glance, Excel might seem to be inappropriate for Key Stage 1 children but it can easily be tailored to meet their needs by changing the font size, the width of the cells and the colour of the data entry.

So why use spreadsheets? Wheeler (2005) refers to the emancipatory element of computers in removing some of the tedious elements of some tasks. In the case of spreadsheets, they remove the lengthy process of graphs and charts drawing. You should ask yourself whether the computer helps the children learn the science. In the case of spreadsheets the main benefits include:

- the speed and ease of drawing graphs and charts – pie charts, for instance, are easy to interpret but difficult to draw without a computer;
- the speed and ease of calculating outcomes (e.g. averages), especially if the teacher devises a template;
- the fact that it is easy for children to enter the data and have the computer automatically calculate the outcome;
- the simplicity of entering data into prepared tables that can be saved, altered and easily compared with others.

Using databases to collate and sort information has several applications in primary science, although it is usually easier to draw graphs using a spreadsheet. However, the branching databases (or dichotomous databases) are important as an extension from sorting and classifying materials and objects. It leads logically on to children constructing their own keys. The instructions on screen guide users through the process of using the branching database, but is also desirable for children to draw out the structure of the key on paper. A branching database for small animals might look something like the one in Figure 10.2.

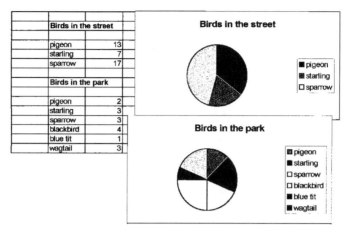

Figure 10.1 What can you learn from these charts? What would children learn?

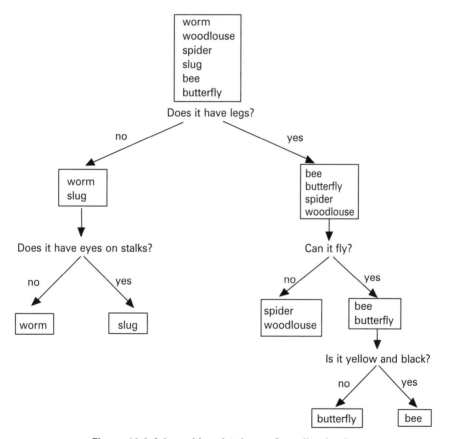

Figure 10.2 A branching database of small animals

Databases such as Microsoft Access are relatively complex to use and rarely contribute much to primary science since usually there are not sufficiently large quantities of information to manipulate. However, a large school-wide database that incorporates all the invertebrate animals and plants in the school grounds, for instance, could be useful if it is added to over the years.

So why use databases? Databases such as those produced specifically for schools by Black Cat and Textease make the manipulation of data easier for children. It is easier to add items to a computer database than it is to add to a paper database. Information can be collated and displayed in a number of ways. The computer can very swiftly identify items requested by the user.

Using sensors to detect and measure changes

Datalogging involves plugging sensors which record temperature, light, sound or movement into a small box attached to the computer. You then run a program that records the readings on the sensors. These readings are then plotted on a graph. The fact that a graph is built up in real time makes this a useful teaching tool. Several manufacturers produce datalogging equipment, the most popular being produced by LogIT and Philip Harris.

Some dataloggers have to be plugged into the computer that records the readings of the sensors. Equipment of this type includes LogIT Live. Other datalogging equipment can be used away from the computer, allowing the data to be uploaded at a later stage. LogIT can be used away from the computer and can record many readings for later use. The change in air temperature outside over a 24-hour period compared with the changes in pond water temperature and the temperature under a large stone in the same period can lead to discussion about why some habitats are more productive than others.

For a practical example of the outdoor use of this approach, see the classroom example at the end of this chapter.

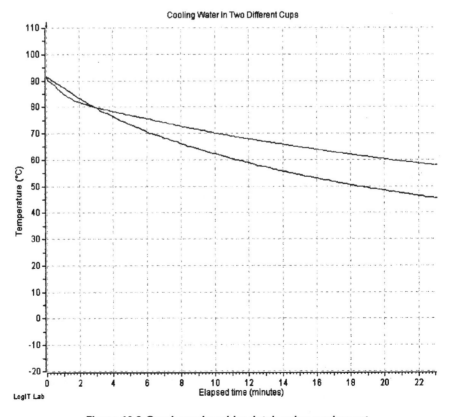

Figure 10.3 Graph produced by datalogging equipment

Here are some of the possible areas of use for datalogging equipment:

Light
● Which sunglasses let most light through?
● Which fabric is best for curtains?
● What happens to the light readings as you move the sensor further away from a torch?
● Which source of light is the brightest?
● Which road safety reflective strip reflects most light?

Sound
● Make a sound plot of the room you are in. Write a sound story to explain the graph.
● Which material makes the best sound insulator?
● What happens to the sound level as you move away from/towards a sound source?

Temperature
● Which insulation material slows cooling best?
● Do large volumes of water cool more slowly than small ones? How does this relate to large and small animals in the Arctic?
● What happens to the temperature of icy water left in the classroom?
● What happens to the temperature in the classroom on a sunny day?

For more on asking questions, see Chapter 7.

Movement sensors
● Do goldfish sleep at night?
● How many birds visit the bird table at each hour of the day?

Pulse sensors
● How does my pulse change when I exercise?
● Which forms of exercise give my heart most exercise?

Higginbotham (2003) shows how datalogging equipment can be used to explore, compare, record, monitor and investigate. These five steps can be used in a variety of contexts ranging from comparing sunglasses to measuring temperature.

Obtaining information from electronic sources

A vast amount of up-to-date information is available on the internet. If children need to find this for themselves, clear training on the most effective ways such as using key words and advanced searches should be given. Much pupil time can be wasted through a vague or inefficient search of internet material, with the risk of children discovering undesirable material even on supposedly protected sites. It is suggested that younger children should be directed to a limited number of specific sites which have been checked by the teacher for suitability, or to material available via the local authority's VLE. A time limit on searching will also help to focus the research. For Key Stage 1 children, it is often better to develop their research skills on the safe and focused environment of a CD-ROM.

RESEARCH SUMMARY RESEARCH SUMMARY RESEARCH SUMMARY **RESEARCH SUMMARY**

Hart (2003) examined sources of information for schools, including Grids for Learning, sources of images and sounds, and websites which deal with science across the world. She also described sources of virtual experiments.

For more on ways of giving instructions, see Chapter 6.

Figure 10.4 Example of activities available on the internet

The internet is changing the way information is gathered. You need to know how best to get the information you require. You may know the address of the information source you are seeking or you can search the web for information on specific science subjects using a variety of search engines.

PRACTICAL TASK PRACTICAL TASK **PRACTICAL TASK** PRACTICAL TASK

The weather is a fascinating resource for teachers. It is easy to find detailed information on a variety of subjects, for example about the changes to the length of the day during the year. For detailed information on sunrise and sunset times, search for 'sunset times UK'. Make notes as to how this could be included in your planning for science.

Using images from digital cameras, film and scanners

Digital cameras and scanners are extremely useful devices for use in primary science. Digital cameras make the results of photography immediately accessible by children and photos can be easily incorporated in their written work. Uses for digital cameras include:

- recording a tree from season to season;
- photographing parts of the school grounds to incorporate into reports emailed to classes in other parts of the country or even across the world;
- recording what the children did at different stages of an experiment;
- recording the changing shape and direction of shadows or clouds;
- using images in children's presentations;
- recording how fruit decays each day;
- recording the growth of seedlings;
- making science material or habitat trails round the school.

Uses for scanners include:

- scanning copyright-free images into the children's work;
- scanning structural details of a flower;
- scanning text from a book for the children to manipulate as part of their work.

Digital cameras are easy to use and the results almost instant. Images can be saved and pasted into a number of documents quickly and at little expense. The use of video webcams is increasing year on year.

Creating illustrations using draw or paint programs

Drawing and paint programs have limited application in science. However, it is relatively easy for children to use drawing programs to draw electrical symbols that can then be duplicated, moved around and pasted into new circuit diagrams. Pictures created in art or ICT sessions using paint or draw programs can be completed as labelled diagrams during a science session, giving purpose to the artwork and the opportunity to improve a drawing which might be difficult with traditional media. Digital movie cameras may be used to record results of experiments which can be replayed for closer examination. They can also be used to present results to an audience with the addition of text and still images.

Communicating through e-mail and networks

The ease and speed of e-mail communication means that it is possible for children in a class in Birmingham to communicate with a similar class in Penzance or Perth, Australia. E-mails can carry attachments including photographs, documents and graphs. This is especially useful when communicating the results of experiments, weather conditions or reports on the environment of the school grounds. The results of plant growth experiments in Aberdeen, London and Sydney in January would be quite different and useful comparisons could be made. Climatic information and information about observations of the Sun and Moon can be sent quickly and easily from other continents. The immediacy of the information from school to school via the internet promotes learning through collaboration and discussion.

Presenting information

Even young children in Key Stage 1 can start to record their data in pictograms and bar charts. The appearance of spreadsheets such as Excel can be adapted through the use of font size and colour and graphics to create child-friendly templates to which children can add data and see the resulting charts created on screen.

When children have completed their investigations, data from a program used to collect and analyse it, such as spreadsheet and imaging programs, can be copied easily into presentation software such as PowerPoint, Hyperstudio or Black Cat presenter. Children can combine their graphs, tables, photos and text in presentations which include motion clips and sound to present a multimedia summary of their learning. A hyperlinked presentation can be browsed by the reader following linked pages on different aspects of one theme. For example, a science-based presentation on the habitats round the school might be browsed by habitat or type of animal or plant found.

RESEARCH SUMMARY RESEARCH SUMMARY RESEARCH SUMMARY **RESEARCH SUMMARY**

O'Connor (2003) reported on how she taught her class of Year 5 children to use PowerPoint. She then asked them to design, author and present their own multimedia slide show focusing on electricity. This required the learners to assemble information, transform and translate the information and draw conclusions. The learners enjoyed the flexibility of approaching the subject matter in different ways, and they enjoyed working collaboratively rather than individually.

Simulation software

Sometimes presented on CD-ROMs, these allow children to explore phenomena which take a long time to occur or to manipulate variables in a virtual experiment which can be repeated and varied many times, quickly and without expense. They help to develop children's ability to think logically and to motivate them by allowing them to take the experiments to extremes which would not normally be possible. Virtual experiments should not totally replace practical experience. The limitations of the software should be discussed with the class.

Interactive programs such as Two-Can's *Polar Lands* and *Electricity and Magnetism* have good ways to extend practical work. In *Electricity and Magnetism*, for example, children can make circuits with the click of a mouse and see what happens when they make parallel and series circuits. They can easily see the effects of switches and resistors. This is not a substitute for handling the real thing but it is a good addition to the children's range of experience. O'Connor (2003) looks in detail at the use of science and specific software such as Interfact and suggests that the use of colourful images grabs the children's attention, in contrast to the use of explanatory text.

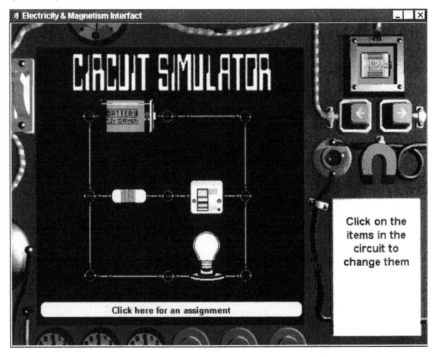

Figure 10.5 Electricity and magnetism circuit simulator

Digital microscopes

Modern digital microscopes allow groups of children to see the complex structures of minibeasts or materials enlarged on a computer or whiteboard screen. Some incorporate time-lapse photographic features, which allow the recording of seed germination and fruit decay. Some allow children to alter the background or the image, such as adding different feature to an invertebrate, which can be motivating and support science learning. They allow images to be saved.

Hart (2003) discusses some uses of a digital microscope. These could include the following.

EYFS

- Fingerprints (print chalky fingerprints on sticky tape to examine).
- Watch a snail eat lunch.
- Compare the properties of rough and smooth objects.

Key Stage 1

- Tadpole movement and development (take photos at each stage).
- Slices of fruit.
- The way different insects move, e.g. woodlice and caterpillars.
- Time-lapse seedling development.
- Time-lapse of melting ice.

Key Stage 2

- Look at teeth soaked in cola.
- Parts of a flower, seeds and leaves.
- Soil samples.
- Time-lapse photography of woodlice in a choice chamber.
- Make a movie of snowflakes melting.
- Video chemical reactions such as vinegar and bicarbonate of soda.

Interactive whiteboards

IWBs enable resources to be shared by the whole class. Children can manipulate diagrams, text, graphs and pictures to show their ideas. IWBs can be used by the teacher to demonstrate skills and model techniques and then by children to try out strategies and predictions. They are useful to demonstrate writing and redrafting of science reports and can be used to motivate children with interactive resources such as experiments and quizzes. Voting systems can allow individual children to express their views and ideas and the teacher to assess their learning.

RESEARCH SUMMARY RESEARCH SUMMARY RESEARCH SUMMARY **RESEARCH SUMMARY**

Earle (2004) examined the way in which IWBs can be used to enhance the teaching of science. She highlighted the fact that text and diagrams can be enlarged, children are motivated to participate and flipcharts can be prepared beforehand. She looked in detail at the way that interactive whiteboards can be used to teach graphing skills. She noted the success of the teaching using IWBs and the benefits of this approach.

IN THE CLASSROOM

In this project about habitats, the teacher used a range of techniques to help her Year 4 class study the school grounds. She used ICT where she felt it was appropriate. The teacher explained that a habitat was where an animal or plant lived. She told the children that they were looking in their school grounds for a range of habitats. They used maps and plans to mark the location of the different habitats they found. They used a digital camera at each location to record what each habitat looked like.

The teacher realised that the children would be particularly interested in the common minibeasts they found in different habitats. At first, the children wanted to find a range of creatures and note their features. They identified them using a range of electronic and paper sources. They also used branching to identify and sort the creatures. The children's information was pooled and a table was completed which showed the names of the habitats and the number of minibeasts found.

Number of each minibeast in each habitat				
	Under the stone	In the short grass	In the long grass	In the flowerbed
spider	1	0	1	1
worm	3	0	0	3
slug	4	0	1	0
snail	1	0	1	1
woodlice	5	0	0	2

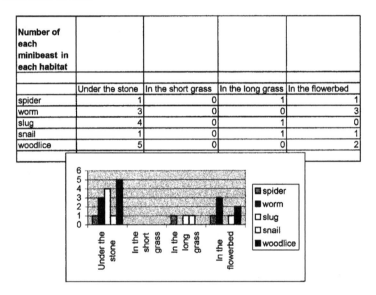

Figure 10.6 Bar chart produced from the data collected

The teacher asked the children why some environments, e.g. under the carpet, under the stone and in the long grass, had so many types of animals while the school field and the soil at the base of a wall had few. She introduced the children to simple datalogging equipment. To show the children how it worked she set up the datalogger to record changes in their own classroom habitat over a 24-hour period.

The children offered their ideas about why some habitats had such a diversity of animal life. Some of the ideas included the suggestion that some habitats did not dry out or heat up during the day as much as others. At first, the children used their fingers to test humidity, thermometers to measure temperature and their eyes to estimate the amount of shade.

The teacher set up a datalogger (which did not need to be connected to a computer) in several different habitats over the course of a week, though she was restricted to daytime because of the chance of theft. She also set up a datalogger to take readings of light, temperature and moisture over the school day. Three contrasting habitats were measured in this way and the results compared.

The teacher had previously communicated with a teacher in another school where the children were also studying their school grounds. The children in the two schools communicated their findings to each other using e-mail attachments, which included the word-processed information, spreadsheets, databases and photographs.

	Minibeasts Key Stage 1	Materials Key Stage 2
Access, select and interpret information	A talking encyclopedia on CD-ROM allows younger children to look up information where previously they could not.	Access up-to-date news using secondary sources like the internet and multimedia software, e.g. to find out about materials used to make a new plane.
Recognise patterns, relationships and behaviours	Use a digital camera to record the features of different habitats. Use a template on a spreadsheet program to graph minibeasts in each habitat. Use branching database to classify animals.	Use datalogging to see which materials let light through and which would make the best blackout blinds. Use branching database to sort materials.
Model, predict and hypothesise	Use a digital microscope with time-lapse photography to see which foods slugs and woodlice prefer.	Use simulation program to test the absorbance or other properties of materials.
Test reliability and accuracy	Use simulation software to: – repeat tests using digital camera to check for accuracy and result; – record results on a spreadsheet for graph averaging.	
Review and modify their work to improve the quality	Use digital film to review test procedures and help identify ways to improve work.	Use word processor and evaluation template to review learning and record it accurately. Review results using graphs with a view to choosing the one most suited to present their data.
Communicate with others and present information	Use word-processing and digital camera or paint program to make a poster of their learning. Use digital graphs to present data. Use slideshow presentation with the above to show work to an audience.	Use a slideshow or kiosk presentation to share learning. E-mail children in another school to compare results.
Evaluate their work	Use reviewing template made by teacher to evaluate learning.	Use online database of other children's work to compare with their own.
Improve efficiency	Use datalogging sensors to record light levels in different habitats or a movement sensor to see if woodlice in a plastic box move around at night. Use clip art to improve posters and labelled diagrams.	Use digital microscope and camera to compare the surfaces of different materials.
Be creative and take risks	Add pictures done in paint program to final report as labelled diagrams.	Use scans and digital photos of materials in the local environment to create a local materials trail.
Gain confidence and independence	Use CD-ROM to search for information in its database, e.g. to find out what minibeasts like to eat.	Use internet to search for information in online databases and articles, e.g. on the strongest and lightest materials.

Table 10.1 How ICT might be used to support the topics Minibeasts and Materials

A SUMMARY OF **KEY POINTS**

> **ICT is a subject in its own right but is also used to support and develop learning in other subjects.**
> **Children can write word-processed reports that include tables, graphs, photographs and drawings.**
> **Spreadsheets and databases can be used to store and manipulate data.**
> **Sensors/dataloggers detect and measure changes in the environment.**
> **Children can obtain information from electronic sources such as websites and CD-ROMs.**
> **Digital cameras, film clips and scanners can be used to provide images.**
> **Draw and paint programs can be used to create illustrations.**
> **Children from different schools and even countries can communicate and exchange work via email, for example to share the result of similar investigations.**
> **Presentation software can be used to present information, including charts, tables, sound and moving images as well as text.**
> **Simulations allow children to explore phenomena that take a long time to occur or to manipulate variables in a virtual environment.**
> **Digital microscopes provide children with opportunities to see enlargements on a computer or IWB screen, including the use of time-lapse photography.**
> **IWBs provide a platform for a wide range of science software.**

M-LEVEL EXTENSION > > > > M-LEVEL EXTENSION > > > >

Using the examples given in Table 10.1, develop ideas for how ICT could be used to support your science teaching and the children's learning in some or all of the following aspects.

- The functioning of organisms: green plants.
- The functioning of organisms: humans and other animals.
- Continuity and change.
- Ecosystems.
- Particle theory and the conservation of mass.
- Electricity and magnetism.
- Energy.
- Forces and motion.
- Light.
- Sound.
- The Earth and beyond.

REFERENCES REFERENCES **REFERENCES** REFERENCES REFERENCES

Earle, S. (2004) Using an interactive whiteboard to improve science specific skills. *Primary Science Review*, 85, 18–20.

Hart, G. (2003) ICT and primary science: learning with or learning from? *Primary Science Review*, 76, 7–8.

O'Connor, L. (2003) ICT and Primary science; learning with or learning from? *Primary Science Review*, 76, 14–15.

Higginbotham, S. (2003) Getting to grips with datalogging. *Primary Science Review*, 9–10.

QCA (2003) *A Scheme of Work for Key Stages 1 and 2: Information Technology*. London: QCA.

Wheeler, S. (2005) *Transforming Primary ICT*. Exeter: Learning Matters.

FURTHER READING FURTHER READING **FURTHER READING** FURTHER READING

Allen, J., Potter, J., Sharp, J. and Turvey, K. (2011) *Achieving QTS: Primary ICT, Knowledge, Understanding and Practice.* Exeter: Learning Matters.

Barber, D., Cooper, L. and Meeson, G. (2007) *Learning and Teaching with IWBs: Primary and Early Years*. Exeter: Learning Matters.

Duffy, J. (2007) *Achieving QTS: Extending Knowledge in Practice: Primary ICT*. Exeter: Learning Matters.

Gillespie, H., Boulton, H., Hramaick, A. J. and Williamson, R. (2007) *Learning and Teaching with virtual learning environments. Learning and Teaching with IWBs: Primary and Early Years*. Exeter: Learning Matters.

11
Health and safety

Professional Standards for the award of QTS

Those awarded QTS must have a secure knowledge and understanding of science that enables them to teach effectively across the age and ability range for which they are trained. To be able to do this in the context of health and safety in science, trainees should:

Q3(a) Be aware of the professional duties of teachers and the statutory framework within which they work.

(b) Be aware of the policies and practices of the workplace and share in collective responsibility for their implementation.

Q21(a) Be aware of current legal requirements, national policies and guidance on the safeguarding and promotion of the well-being of children and young people.

(b) Know how to identify and support children and young people whose progress, development or well-being is affected by changes or difficulties in their personal circumstances, and when to refer them to colleagues for specialist support.

Q30 Establish a purposeful and safe learning environment conducive to learning and identify opportunities for learners to learn in and out of school contexts.

Q32 Work as a team member and identify opportunities for working with colleagues, sharing the development of effective practice with them.

The guidance accompanying these standards clarifies these requirements and you will find it helpful to read through the appropriate section of this guidance for further support.

Introduction

Health and safety requirements vary according to the science being taught, the resources and methods used, the number of children involved, the level of supervision required, the physical layout of the classroom or teaching area, and so on. They can also vary from school to school and authority to authority. All teachers should be familiar with the layout of the school, where entrances and emergency exits are, the location of fire alarms and firefighting appliances, what to do in the case of fire drills and other emergency procedures, and the siting of telephones, particularly if engaged in activities associated with running an after-school science club.

Legal requirements

Teachers, including trainees and NQTs, have a common-law duty to ensure that children are safe on school premises and during off-site visits. This is bestowed on them by acting *in loco parentis* ('in the place of a parent'). When acting *in loco parentis*, teachers are generally expected to exercise more care than would normally be expected of a careful parent. All teachers, including trainees and NQTs, also have a duty to work in accordance with the Health and Safety at Work etc. Act 1974. The Act points out that employers, including local authorities (LAs) and governing bodies, are required to keep schools as safe as is reasonably practicable and foreseeable for all who use or visit them. Sections 7 and 8 of the Act are more specific:

- teachers have a statutory and contractual obligation to co-operate with their employers in all matters relating to health and safety;
- teachers must not only take reasonable care for the health and safety of themselves, they must also take reasonable care for the health and safety of others;
- teachers should not intentionally or recklessly interfere with or misuse anything provided in the interests of health and safety.

Just as acting *in loco parentis* and the Health and Safety at Work etc. Act are important to schools as a whole, they also apply when teaching science. Enquiries about health and safety in any school should be directed to the appointed health and safety representative or, for primary science, to the science subject leader.

The language of health and safety includes hazards, risks, risk assessment and risk management:

- *hazards* – result in exposure or vulnerability to injury, may be connected with the resources and methods used by teachers or children in teaching and learning science;
- *risks* – the perceived or actual likelihood of hazards resulting in injury, may be high or low depending on the nature of the hazards, the injury likely to be sustained, and the number of people likely to be involved;
- *risk assessment* – identifies all hazards and likely outcomes in advance of their happening, provides the basis for selecting safe ways to proceed;
- *risk management* – identifies and adopts strategies to produce a zero risk level or at least minimise the potential for risk, ensures that all reasonable precautions are taken and codes of conduct followed.

RESEARCH SUMMARY RESEARCH SUMMARY RESEARCH SUMMARY **RESEARCH SUMMARY**

Codes of conduct for health and safety in primary science are available in all school and LA science policies. These should be followed rigorously. The Association for Science Education (ASE – www.ase.org.uk), the School Science Service (CLEAPSS – www.cleapss.org.uk) and the Scottish Schools Equipment Research Centre (SSERC – www.sserc.org.uk) also provide health and safety information for primary science. The ASE's publication *Be Safe!* (Abbott, 2001) and health and safety guides and newsletters from CLEAPSS are particularly useful and should be consulted as a matter of routine. Additional advice about keeping animals in schools can be obtained from the Royal Society for the Prevention of Cruelty to Animals (RSPCA – www.rspca.org.uk). All schools and LAs issue codes of conduct for off-site visits. These should be followed rigorously.

Safe science

At all times, teachers are advised to avoid any potential for allegations of negligence associated with work in science and never to assume that any activities that get into print are safe. The following pointers draw attention to some of the more common aspects of health and safety requiring careful attention.

Humans

- Avoid touching, tasting, handling or picking up foodstuffs unless it is absolutely safe to do so.
- Ensure meticulous care in choice, storage, preparation, use and disposal of foodstuffs.
- Avoid contamination of foodstuffs.
- Sterilise all utensils and work areas.
- Ensure meticulous attention to personal hygiene.
- Use disposable gloves.
- Ensure that ALL teeth and bones are clean and sterilised before use.
- Be aware that the use of smoking machines is not always allowed.
- Avoid exercises which unreasonably test stamina or strength.
- Be aware of self-image and other emotional issues, including using family trees or collecting information on weight and height.
- Be aware of ethical issues in relation to human participants, clothing, certain foodstuffs and animals.
- Be aware of sensitivities, including about disabilities.

Other animals

- Ensure that animals are selected for the classroom with great care.
- Ensure that specialist advice is sought or reputable texts are consulted.
- Ensure that only reputable animal suppliers are used.
- Avoid animals that transmit diseases to humans, are difficult to keep, or cause infestation.
- Be aware that not all animals are safe to handle.
- Avoid touching animals unless it is absolutely safe to do so.
- Be aware of allergies and reactions to certain animals.
- Ensure meticulous attention to personal hygiene.
- Use disposable gloves.
- Never keep any wild animals brought in by children.
- Be aware of laws governing collecting or disturbing certain rare or endangered and protected species.
- Do not keep mammals or birds in rooms where humans work regularly.
- Ensure that cages and aquariums are cleaned regularly.
- Restrict the free movement and access of animals.
- Prevent contamination of foodstuffs.
- Ensure the care, safety and well-being of animals, especially during weekends and longer breaks.
- Treat bites and stings immediately.
- Avoid working in areas with contaminated soils (e.g. by glass, faeces, fertilisers).
- Ensure that all animals collected from the wild are returned.
- Be aware of sensitivities involving animal experimentation.

Green plants

- Be aware that not all plants are safe to handle.
- Be aware that some wild flowers are protected or endangered and should not be picked at all.
- Avoid touching, tasting or eating plants unless it is absolutely safe to do so (including fruits and seeds).
- Avoid plants which sting or bear thorns.
- Be aware of allergies and reactions to certain plants.
- Avoid working in areas with contaminated soils (e.g. by glass, faeces, fertilisers).
- Ensure meticulous attention to personal hygiene.
- Use disposable gloves.

Living things in their environment

See also comments on *humans*, *other animals* and *green plants* above.

- Avoid unnecessary exposure to the Sun.
- Avoid unnecessary exposure to harsh weather conditions.
- Avoid working in areas with contaminated soils (e.g. glass, faeces, fertilisers).
- Avoid touching, tasting or eating plants unless it is absolutely safe to do so (including fruits and seeds).
- Avoid touching animals unless it is absolutely safe to do so.
- Ensure sensible precautions when conducting litter surveys.
- Ensure meticulous attention to personal hygiene.
- Use disposable gloves.
- Ensure only safe sources of micro-organisms are used and are disposed of safely.

Materials

- Be aware of and avoid flammable and toxic materials and materials which cause irritation.
- Be aware of and ensure that only approved acids and alkalis are used.
- Avoid using very fine powders and materials which produce fine dusts when cut.
- Avoid superglues and adhesives which give off strong fumes.
- Avoid using boiling water and steam directly from kettles.
- Avoid gases other than air or the carbon dioxide in fizzy drinks.
- Burn only small amounts of materials under strict supervision.
- Avoid naked flames other than using candles in sand trays.
- Use cookers, hot plates, microwaves, fridges and freezers under strict supervision.

Electricity and magnetism

- Ensure that all mains appliances are thoroughly and regularly checked.
- Be aware of the dangers of using mains electricity.
- Be aware that not all batteries are safe to use in school.
- Avoid cutting batteries open.
- Ensure that loose iron filings are stored securely and only used in sealed containers.
- Be aware that small magnets could be swallowed or placed in ears and noses.
- Ensure that electromagnets do not overheat.

Forces and motion

- Ensure that fingers and toes are kept well away from heavy objects likely to fall on or roll over

Figure 11.1 How many health and safety violations can you find? Cartoons like this can be used to help children identify hazards and risks and to carry out primary science safely.

them when released down ramps or in free-fall.
- Ensure eye protection is worn when stretching springs and elastic bands or when twisting and bending brittle materials.
- Avoid climbing on chairs and tables when releasing 'helicopters' and other flying objects.
- Avoid reckless throwing of paper darts and gliders.
- Be aware of breathing difficulties if playing games like 'blow football'.

Light and sound

- Protect eyes at all times.
- Avoid shining bright lights into eyes.
- Avoid using mirrors and lenses with rough or sharp edges.
- Avoid foreign objects from entering eyes.
- Protect ears at all times.
- Avoid listening to loud sounds for any length of time.
- Avoid foreign objects from entering ears.

The Earth and beyond

- Avoid looking directly at the Sun at all times and *never* look directly at the Sun through binoculars or a telescope.
- Supervise all indirect observations of the Sun.
- Avoid observing the Moon through binoculars or a telescope for long periods of time.
- Be aware that some models of the Solar System and the Earth–Sun–Moon system contain small parts which could be swallowed or placed in ears and noses.
- Ensure that parents are informed of homework activities which require observations of the night sky.

PRACTICAL TASK PRACTICAL TASK **PRACTICAL TASK** PRACTICAL TASK

Make sure that you have consulted all the sources of information provided here before you next teach science (see **Research summary**, **Further reading** and **References**). Where appropriate, pay particular attention to the legal requirements for keeping living things in the classroom. Demonstrate that you have considered health and safety by making a clear health and safety statement in your planning.

Using computers

- Ensure safe and comfortable working conditions at all times (select suitable benches and chairs with height, reach and posture in mind).
- Store cables and other loose wiring safely away from feet.
- Avoid placing food and drink near computers or other electrical appliances.
- Be aware of neck strain, eye strain and repetitive strain injuries when using computers for long periods of time.
- Avoid excessive screen glare and reflection.
- Be aware of crowding around computers.

THE BIGGER PICTURE THE BIGGER PICTURE THE BIGGER PICTURE

Just as you must protect and safeguard the children in your care, so you must also take care over your own privacy. When 'off duty', you may want to keep in touch with friends using social networking sites. Be careful what information you share on these, including any photographs of your social life. Remember that children and parents may look you up! It is not appropriate to accept them as friends on such sites. It is important to follow all relevant guidelines on communicating with children and their parents/carers, for example on using your own personal mobile, including those issued by your school, the local authority or nationally. The Child Exploitation and Online Protection (CEOP) Centre is a useful resource for guidance on cyberbullying, and the safe use of social networking sites and the internet. It has materials suitable for children of different ages and advice for parents and carers and for professionals working with children and young people. See their website (www.ceop.police.uk) for further details.

Risk assessment

Borrows (2003) suggests that a risk assessment is the key to sensible safety. Ask yourself, *How likely is it that things will go wrong?*, *How serious would it be if they did go wrong?*

Off-site visits

Off-site visits involve children and teachers in travelling and working beyond the school grounds. Off-site visits therefore include day trips to local parks and ponds, museums and science centres, zoos, botanical gardens, wildlife centres and beaches. They also include residential trips to environmental and other field study centres. Catling (2006) explains how to balance the excitement of working outdoors with possible hazards. Off-site visits require special attention.

- Nominated leaders should be suitably trained, qualified and experienced.
- Nominated leaders, together with all accompanying teachers, are ultimately responsible and accountable for everything that occurs.
- Only approved transportation operators should be used (e.g. minibus, coach, train, underground).
- Only approved locations and environmental and other field study centres should be used.
- A thorough risk assessment should always be carried out in advance of the trip and following a reconnaissance visit to the locations involved.
- Proper consultation should always take place with headteachers and, if appropriate, governing bodies and the LA.
- Proper consultation should always take place with parents and guardians and prior permission should always be obtained in writing.
- Communication between the travelling party and the school should be possible at all times (e.g. by mobile phone).
- All codes of conduct offered by schools, LAs and other responsible bodies should be followed rigorously.

IN THE CLASSROOM

Mrs Evans and Mr Davies were getting ready to take their classes to the city museum for the last of several visits while teaching about humans and other animals. The museum was five miles away and the children were carried there and back by an approved coach operator. Leaving the school, Mr D let Mrs E know that he had counted 59 children on board. The visit passed by without incident, as usual, and Mrs E counted 59 children back onto the coach. Both teachers had been back at the school for about 15 minutes when the school secretary came running into the staff room to tell them that they'd left a child behind! But how had this happened? How it had happened was easy: Mr D had miscounted. There were, of course, 60 children present. Both teachers should have counted independently, probably twice, and cross-checked. Mrs E, thinking that all were back on board, instructed the driver to go. While counting independently and cross-checking would not have revealed the missing child, had both teachers miscounted initially, a simple check around the museum galleries would have revealed their mistake. Both teachers were at fault. Fortunately for them, the child was 'retrieved' safely, praised for his outstanding interest in science, and the matter passed without further incident. Some schools assign children to specific seats and keep a seating plan readily available on such trips. This is a much better strategy than counting.

EMBEDDING ICT EMBEDDING ICT **EMBEDDING ICT** EMBEDDING ICT

Many useful local authority policies are available online, on the intranet via the LA's portal. This may include the policy and procedures for arranging off-site visits, especially those involving outdoor and adventurous activities or those that include water, for example visits to coastal, lakeside or riverside habitats. Make sure that you familiarise yourself with all important policies relating to health and safety.

First-aid and administration of medication

No teacher is obliged to become a first-aider or to act as an appointed first-aider within a school. All teachers should, however, know who these people are, where they can be found when needed, and where the nearest first-aid box is. All teachers should be familiar with school rules and procedures for dealing with and reporting accidents requiring first-aid. Only appointed first-aiders are usually responsible for:

- maintaining and updating first-aid boxes;
- taking charge of emergency situations;
- administering emergency first-aid where they have been trained and feel confident and competent to do so.

At the time of a serious accident, one which requires prompt and immediate action, teachers may be required to deal with the emergency while waiting for a first-aider or other help to arrive. Serious accidents might include:

- burns and scalds;
- cuts and bleeding;
- bites and scratches;
- spillages of chemicals onto skin;
- splashes of chemicals into eyes;
- ingestion of certain food substances, chemicals or poisons;
- choking and other breathing difficulties;
- electric shock and electrocution;
- loss of consciousness.

The St John Ambulance publication *Emergency Aid in Schools* (2006) does provide help and St John Ambulance will advise on suitable courses of first-aid training. Just as no teacher is obliged to become a first-aider or an appointed first-aider within a school, no teacher is obliged to administer medicines or to supervise children taking them, particularly if the timing of administration is crucial to a child's health or where medical and technical knowledge is required. All teachers should, however, know which of their children require medication to be administered during the school day, the properly qualified, trained and appointed individuals within the school responsible for undertaking such tasks, and the school's procedures for dealing with medical emergencies. All teachers should be familiar with school rules and regulations for receiving, handling and storing medicines. Any of the teachers' unions will provide clear guidance on all of these issues.

IN THE CLASSROOM

Chris had been swinging backwards and forwards on his chair all morning. His teacher, working with a group investigating forces at another table, had told him twice already to stop and what might happen if he didn't. Eventually Chris did fall over. In fact, he disappeared from view with a bit of a thud! 'Are you going to lie there all day Chris, or are you going to join the rest of us and do some work?' his teacher called out. 'He's bleeding sir!' came a shout from the girl sitting next to him. What would you have done? Fortunately, Chris's teacher knew exactly what to do. Noting the large cut above Chris's eye, and the amount of bleeding involved (Chris had hit his head on the edge of the desk as he went down), the class were settled quickly into their seats. Two children were sent to get the deputy headteacher immediately. After hastily putting on the protective gloves kept in the teacher's desk, direct pressure was applied to Chris's wound to help stop the bleeding. Once the deputy headteacher arrived and took over, Chris's teacher left the room to arrange for Chris to visit the local health centre and for Chris's parents to be informed. The school caretaker was called in to deal with the blood on the classroom floor. The whole incident was dealt with in minutes. Chris returned to school later that day. The incident was used to illustrate how easily accidents happen without due care and attention and to praise the children for their calm and appropriate behaviour.

Teaching about health and safety

Teaching about health and safety in primary science should be looked on as a routine part of teaching science as a whole. Useful teaching strategies include:

- effective questioning (e.g. 'What do you think will happen if...?', 'Who can tell me...?', 'Why do you think...?');
- instruction, demonstration and explanation (e.g. safe ways to proceed, what to do and not to do and why);
- display (e.g. health and safety symbols, children's posters and formal science reports, written and pictorial reminders);
- classroom rules and procedures (e.g. noise levels, movement around the classroom, appropriate behaviour, accident and emergency drills);
- example (e.g. practise what you preach).

Children's knowledge and understanding of the importance of health and safety should progress throughout the primary years. By the end of Key Stage 1, children should be able to recognise that there are hazards and dangers associated with certain aspects of primary science (scientific enquiry, life processes and living things, materials and their properties, and physical processes) and, with help from their teachers, begin to identify and assess those hazards and dangers for themselves. They should also be increasingly aware of how to care for and look after each other and other living things. By the end of Key Stage 2, children should be able to routinely recognise hazards and dangers associated with certain aspects of primary science and to identify and assess those hazards and dangers for themselves. They should be encouraged and able to apply their existing knowledge and understanding of health and safety to new situations and respond accordingly.

Figure 11. 2 Safety signs for classroom use. Children could explore designing their own and choosing suitable colour combinations.

A SUMMARY OF **KEY POINTS**

> Primary schools are relatively safe places and reported accidents or incidents as a result of primary science activity are rare.

> Teachers are acting *in loco parentis* and are responsible for the safety of the children in their care, their own health and safety and that of others in school.

> Risk assessment and risk management can help to avoid preventable injury.

> Teachers, including trainees and NQTs, should be familiar with codes of conduct for health and safety in primary science, including those for off-site visits, for reporting accidents or incidents requiring first-aid, and for dealing with other medical emergencies.

> On matters of health and safety there is no substitute for proper training, qualifications and experience.

> Teaching about health and safety in primary science should be looked on as a routine part of teaching science as a whole.

M-LEVEL EXTENSION > > > > M-LEVEL EXTENSION > > > >

Consider whether any of the health and safety recommendations provided in this chapter, including in the classroom examples and research summaries, and by the reference texts and further reading suggestions, apply more widely than in science. What other subjects are likely to need similar levels of attention to the safeguarding of pupils?

REFERENCES REFERENCES **REFERENCES** REFERENCES REFERENCES

Abbott, C. (ed.) (2001) *Be Safe! Some Aspects of Safety in School Science and Technology for Key Stages 1 and 2*. Hatfield: Association for Science Education.

Borrows, P. (2003) Managing health and safety in primary science. *Primary Science Review*, 79, 18–20.

Catling, S. (2006) Planning for learning outside the classroom, in Arthur, J. (ed.) *Learning to Teach in the Primary School*. Oxford: Routledge.

St John Ambulance (2006) *Emergency Aid in Schools*. London: Order of St John.

FURTHER READING FURTHER READING **FURTHER READING** FURTHER READING

Excellent sources of relevant materials discussing essential aspects of health and safety can be found via the following websites:

The Association for Science Education: www.ase.org.uk

The Health and Safety Executive: www.hse.gov.uk

The Royal Society for the Prevention of Cruelty to Animals: www.rspca.org.uk

The School Science Service: www.cleapss.org.uk

The Scottish Schools Equipment Research Centre: www.sserc.org.uk

analogy A description or physical representation of something which behaves in a similar way to a science concept, enabling the learner to gain a better understanding of the concept, for example the analogy of current flow in a circuit to the flow of water in pipes.

buffer box A box which has sensors plugged into it. In turn, the buffer box is plugged into the computer.

CD-ROM A disc which contains information or programs.

closed question A question to which the range of answers is limited – often to a single, correct answer.

concept mapping A diagram in which science concepts or ideas, often in text form, are linked to show their relationships. Concept maps can be used to reveal a learner's understanding within a wide conceptual area such as 'light'.

conceptual understanding An understanding of scientific ideas, for example force.

criterion referenced Related to specific behaviour or benchmarks.

database A program which collates data.

datalogger The kit required to sense the environment (for example temperature, sound or light), which includes a buffer box, sensor and software.

diagnostic Assessment through which difficulties can be recognised in order to inform the next teaching step. This is distinct from formative assessment in that diagnostic assessment is usually done by an expert in a particular field (for example profound learning difficulties) with the process taking considerable time.

differentiation Provision made by the teacher to enable children of different abilities or needs to learn effectively.

elicitation Techniques used to find out what children know and understand about a scientific idea or concept.

evaluative Assessment which is planned to evaluate the success of some teaching. This form of assessment could be used, for instance, to check whether one form of teaching is more effective than another.

evidence Information gathered to support or disprove an idea.

Excel A proprietary spreadsheet program for a PC.

formal assessment Planned for and sometimes recorded.

formative Assessment to recognise the child's achievements and difficulties to inform the next teaching steps. This is distinct from diagnostic testing in that teachers normally carry out formative assessment during ordinary teaching activities.

generalisation A statement which describes a series of observations which have something in common, for example most metals conduct electricity well.

Google One of many search engines.

hypothesis An explanation that leads to a prediction which can be tested. It draws on some scientific knowledge. A prediction may be based on previous experience or reasoned expectations without drawing on scientific knowledge or being phrased in a way that can be tested.

illustrative activities Practical activities in which children are guided towards under-standing a particular scientific idea.

informal assessment Not planned for and generally unrecorded.

internet A system of computers linked by telephone lines.

iterative Comparing with child's previous performance.

keys Systematic ways of identifying animals or plants. The simplest binomial keys have questions to which the answer is either yes or no, so that answering the questions in turn leads to the correct name for whatever has to be identified. For example, a key for identifying trees by their leaves might ask, 'Is the leaf made up of smaller leaves?' If the answer is 'no' that leads to a further question which applies to simple leaves, then to another and so on until it is identified. Biological keys are the commonest but they can be used to identify materials. Questions have to be carefully worded. Children can understand keys better if they first make their own for a few objects and then learn to use published keys with increasing sophistication. Computers are useful for making their own and for using existing keys on software.

learning intentions or objectives A description of the learning the teacher expects the children to achieve.

learning outcomes A description of the learning the children are expected to demonstrate in order to show success in the learning objectives.

lesson evaluation A reflective and critical account of a teaching unit, which highlights both the children's and the teachers' responses and sets out targets for the future.

lesson pace The rate at which the various learning experiences change within a single lesson.

lesson structure The form that a lesson takes in terms of timing and features such as lesson introduction, development and conclusion.

long-term planning An overview of the learning programmes for a whole school over the period of a complete year or phase – often described as a scheme of work.

medium-term planning A broad plan of the learning programme for typically a half term or whole term – often described as a topic plan. Some schemes of work include detailed medium-term planning.

model Something which represents and behaves similarly to a phenomenon in the real world, for example a picture of the molecules in a solid, liquid or gas.

multimedia A computer program which uses text, sound and images.

norm referenced Children measured against others.

open question A question to which there is a wide range of acceptable answers.

practical activity A learning situation in which children have some hands-on experience using physical equipment.

predict Make an informed statement about something which will happen in the future.

procedural understanding An understanding of the ways in which science works, and the processes and methods used in science. This involves bringing together the skills that will be needed to carry out an investigation with an understanding of the procedures such as designing appropriate investigations, deciding what measurements to take, how to present and interpret data, and whether they are valid.

progression Provision which will enable a child to move forward in understanding within a subject.

reliability The degree of trust we should place in the data – i.e. whether the experiment is likely to yield similar results if repeated.

restructuring The reforming of science concepts as a result of a learning experience.

risk assessment Identifies all hazards and likely outcomes in advance of their happening, providing a basis for selecting safe ways to proceed.

scheme of work See long-term and medium-term plans.

science The study of how and why everything in the Universe behaves the way it does.

science concept A 'big idea' in science such as force or the particle theory of matter.

science enquiry A learning experience in which the child takes some responsibility for planning, executing and reflecting on an investigation of a science phenomenon.

scientific community Scientists around the world who communicate with each other, constructively criticise each other's findings and share some common beliefs about science concepts.

scientific skills A range of abilities which enable people to behave scientifically.

scientific theory A description made to explain a collection of scientific observations which can then be subjected to scrutiny by the scientific community and matched against a range of evidence which might support or disprove it.

search engine A site on the internet which searches for information which you request.

secondary sources Information which is gained, not through first-hand experience but through the use of sources such as books, ICT or discussions with others.

short-term planning Lesson planning.

SI units Standard international units of measurement and abbreviations are used in science, for example newtons (N) are the unit for measuring forces.

spreadsheet A program which allows data to be entered into a table. The data can then be used in calculations or the production of graphs.

standardised Related to a set of generally accepted criteria.

summative Where assessment is recorded systematically with the aim of establishing markers showing children's attainment. This is usually done at the end of a key stage or at the end of a year.

teacher exposition The provision of information which is provided directly by the teacher orally, often within the introduction to a lesson.

teaching objectives The objectives which a teacher has for a learning programme such as a single lesson. These are often set out in terms of the intended learning objectives for the children.

technology The application of scientific ideas, as well as those from other disciplines, in order to solve a human problem or need.

template A document created using either a spreadsheet or word processor which is saved as a template (on PCs) or stationery (Macs). If opened and altered, the changed version is saved, leaving the original template unaltered.

validity Whether an investigation answers the question. In an experiment, procedures such as controlling variables are used so the data is more likely to be valid.

variables: *Continuous* when they can have any value on a scale (for example, temperature or length).
Discrete when they can only have whole number values (for example, the number of layers of tissue paper needed to stop light shining through).
Categoric when we just assign things to a particular category (for example, boys and girls; eye colours).

variation Between repeated measurements of the same thing may derive from unexpected differences in what is being investigated or from errors. Because it is difficult to do exactly the same on every test, we get random experimental error, which may make our measurement larger or smaller. We can attempt to reduce this by taking an average of several measurements, but averaging will not compensate for a systematic error that always affects our measurement in the same direction.

webcam A camera connected up to the net directly and for long periods, allowing distant observers to see what is happening.